T0258791

Structural Dynamic Systems Computational Techniques and Optimization

Techniques in Buildings and Bridges

Gordon and Breach International Series
in Engineering, Technology and Applied Science

Volumes 7–15

Edited by Cornelius T. Leondes

Books on **Structural Dynamic Systems Computational
Techniques and Optimization**

Volume 7 *Computer-Aided Design and Engineering*

Volume 8 *Finite Element Analysis (FEA) Techniques*

Volume 9 *Optimization Techniques*

Volume 10 *Reliability and Damage Tolerance*

Volume 11 *Techniques in Buildings and Bridges*

Volume 12 *Seismic Techniques*

Volume 13 *Parameters*

Volume 14 *Dynamic Analysis and Control Techniques*

Volume 15 *Nonlinear Techniques*

Previously published in this series were volumes 1–6 on
Medical Imaging Systems Techniques and Applications

Forthcoming in the *Gordon and Breach International Series
in Engineering, Technology and Applied Science*

Mechatronics Systems Techniques and Applications

Biomechanical Systems Techniques and Applications

Computer-Aided and Integrated Manufacturing Systems Techniques
and Applications

Expert Systems Techniques and Applications

This book is part of a series. The publisher will accept continuation orders which may be
cancelled at any time and which provide for automatic billing and shipping of each title in the
series upon publication. Please write for details.

Structural Dynamic Systems Computational Techniques and Optimization

Techniques in Buildings and Bridges

Edited by

Cornelius T. Leondes

Professor Emeritus
University of California
at Los Angeles

Gordon and Breach Science Publishers

Australia • Canada • China • France • Germany • India •
Japan • Luxembourg • Malaysia • The Netherlands •
Russia • Singapore • Switzerland

Amsteldijk 166
1st Floor
1079 LH Amsterdam
The Netherlands

British Library Cataloguing in Publication Data

Structural dynamic systems computational techniques and
    optimization : techniques in buildings and bridges. –
    (Gordon and Breach international series in engineering,
    technology and applied science ; v. 11 – ISSN 1026-0277)
    1. Structural dynamics 2. Bridges - Design and construction
    3. Buildings - Design and construction 4. Computer-aided
    engineering
    I. Leondes, Cornelius T.
    624.1'76

ISBN 90-5699-655-X

CONTENTS

SERIES DESCRIPTION AND MOTIVATION

Many aspects of explosively growing technology are difficult or essentially impossible for one author to treat in an adequately comprehensive manner. Spectacular technological growth is made stunningly manifest by any number of examples, but, just to note one here, the Intel 486 IBM-compatible PC was first introduced in late 1989. At that time the price of this PC was in the $10,000 range and it was thought to be much too powerful for widespread use. By early 1992, a little more than two years later, the price had dropped to $1,000 and it was felt that much more power was needed, leading directly to the Pentium IBM-compatible PC. A similar price reduction pattern has already been projected for the Pentium computer, and, in view of the recent history of the 486, it is difficult to suggest that the same "power hungry" pattern will not occur again in a similar time span. The Pentium is presently planned as a 1,000-MHz processor to be called the Flagstaff in the year 2000. The CD-ROM is presently evolving to the DVD (Digital Versatile Disk) with data storage capability of a greater order of magnitude. A DVD-ROM can hold a database of all the phone numbers and addresses in the United States, which would normally require multiple CD-ROMs. And the DVD format has room to grow. In any event, these examples and their clear implications with respect to the many application-oriented issues in diverse fields of engineering, technology and applied science and their continuing advances make it obvious that this series will fill an essential role in numerous ways for individuals and organizations.

Areas of major significance will be defined and world-class co-authors identified as contributors for essential volumes in respective areas. These areas will be determined by criteria including:

1. Will volumes fill important textbook voids in respective areas?

2. In some cases, a "time void" for an important area will clearly suggest the need for a volume. For example, the important area of Expert Systems might have a textbook void of several years that "requires" an important new volume.

3. Are these technology areas that simply cannot sensibly be treated comprehensively by a single author or even several co-authors?

Examples of areas requiring important volumes will be carefully defined and structured and might include, as the case arises, volumes in:

1. Medical imaging

2. Mechatronics

3. Computer network techniques and applications

4. Multimedia techniques and applications

5. CAD/CAE (Computer Aided Design/Computer Aided Engineering)

6. FEA (Finite Element Analysis) techniques and applications

7. Computational techniques in structural dynamic systems

8. Neural networks (as might possibly be suggested by a significant time void in the textbook literature)

9. Expert Systems (again, depending on a possible significant time void).

One of the most important aspects of this series will be that, despite rapid advances in technology, respective volumes will be defined and structured to constitute works of indefinite or "lasting" reference interest.

SERIES PREFACE

The first industrial revolution, with its roots in James Watt's steam engine and its various applications to modes of transportation, manufacturing and other areas, introduced to mankind novel ways of working and living, thus becoming one of the chief determinants of our present way of life.

The second industrial revolution, with its roots in modern computer technology and integrated electronics technology — particularly VLSI (Very Large Scale Integrated) electronics technology, has also resulted in advances of enormous significance in all areas of modern activity, with great economic impact as well.

Some of the areas of modern activity created by this revolution are: medical imaging, structural dynamic systems, mechatronics, biomechanics, computer-aided and integrated manufacturing systems, applications of expert and knowledge-based systems, and so on. Documentation of these areas well exceeds the capabilities of any one or even several individuals, and it is quite evident that single-volume treatments — whose intent would be to provide practitioners with useful reference sources — while useful, would generally be rather limited.

It is the intent of this series to provide comprehensive multi-volume treatments of areas of significant importance, both the above-mentioned and others. In all cases, contributors to these volumes will be individuals who have made notable contributions in their respective fields. Every attempt will be made to make each book self-contained, thus enhancing its usefulness to practitioners in a specific area or related areas. Each multi-volume treatment will constitute a well-integrated but distinctly titled set of volumes. In summary, it is the goal of the respective sets of volumes in this series to provide an essential service to the many individuals on the international scene who are deeply involved in contributing to significant advances in the second industrial revolution.

PREFACE

Structural Dynamic Systems Computational Techniques and Optimization

Techniques in Buildings and Bridges

Many high-rise buildings have been constructed in recent decades, particularly in Western Pacific Rim countries. Some due for completion soon and others to be completed in the not-too-distant future include:

1. Grand Gateway at Xu Hui, Shanghai

2. Tomorrow Square, Shanghai

3. Xiamen Post and Telecommunications Building, Xiamen, China

4. Petronas Twin Towers, Kuala Lumpur, Malaysia

5. Shanghai World Financial Center.

Petronas Twin Towers (1,483 ft. in height) and the Shanghai World Financial Center (1,509 ft. high) compare with Chicago's modernist block-glass Sears Tower (1,450 ft.). While these represent the ultimate challenges in structural, as well as foundational, aspects of buildings, there are challenges for structural aspects of buildings of various heights. These are driven by desire for the most economical structure that at the same time gives attention to seismic resistance, lighter self-weight with its direct implications in lower foundation costs, reduced structural construction time, and other significant factors. These trends have implications in more complex structural behavior, the requirement to predict behavior of structures with adequate accuracy, and numerous other aspects. Trends in modern bridge structures are similar. This book is a comprehensive treatment of the issues and advanced techniques involved in building and bridge structures. Numerous illustrative examples are included.

This is the fifth volume in the set of 9 volumes on structural dynamic systems. Subjects treated are:

1. Computer-Aided Design and Engineering

2. Finite Element Analysis (FEA) Techniques

3. Optimization Techniques

4. Reliability and Damage Tolerance

5. Techniques in Buildings and Bridges

6. Seismic Techniques

7. Parameters

8. Dynamic Analysis and Control Techniques

9. Nonlinear Techniques.

In the first chapter of this volume, Goman Wai-Ming Ho and Siu-Lai Chan tell us steel-framed structures are the natural choice for buildings of relatively modest height as well as high-rise skyscrapers because they offer shorter construction time, better seismic resistance, lighter self-weight, which in turn implies lower foundation costs, and, not surprisingly, as a result, are the generally adopted solution. With the trend toward constructing slender steel structures, structural behavior is becoming more complex. Hence, rigorous methods have been developed and continue to be developed for analysis of such structures. This chapter comprehensively treats pertinent and related issues, and many illustrative examples are included.

According to Chui and Chan (chapter 2), in conventional analysis and design of steel structures, beam-column joints are usually assumed to be either perfectly rigid or ideally pinned for simplicity. This implies the angle between adjoining members remains unchanged for rigid-joint case and no moment will be transferred between adjoining members for pinned-joint case. However, in reality, experiments demonstrate that connections behave nonlinearly in a manner between the two extreme cases. At connections, a finite degree of joint flexibility exists; connections in practice are, therefore, semi-rigid. The influence of semi-rigid connections on realistic structural response has been recognized and provision for semi-rigid connections has been given in some national steel design codes such as American Load and Resistance Factored Design (LRFD) (1986), British Standard BS5950 (1985), and Eurocode 3 (1988). These issues and others are essential in determining transient analysis of semi-rigid steel frames characteristic of structural steel-frame buildings.

In chapter 3, Chang, Yang and Swei discuss the concept of structural systems control and the increasing interest and development it has been receiving. National and international panels composed of leading figures have been formed to determine study directions and developments, and international conferences have been held on this broad area. Such structural control techniques have a number of goals, including control of structures subjected to earthquakes with a view toward reducing their structural effects, control of tall buildings exposed to random wind loads, and other applications. In general, an active structural control system consists of: (a) sensors installed at suitable locations to measure either external disturbances or system responses, or both; (b) devices to process measured information and compute necessary control forces based on a given control algorithm; and (c) actuators to produce required active control forces. Active control forces are determined by measured information provided from sensors, as well as the particular control algorithm used.

The advent of tubular systems has been a significant development in modern building structural systems (chapter 4). Most of the world's tallest buildings, of both steel and concrete, have tubular systems as the basic load resisting system. This system not only provides high structural efficiency but also large column-free space, suitable for flexible usage. Singh and Nagpal tell us that in this type of system, most of the load-carrying material is on the periphery in the form of closely spaced columns interconnected by deep spandrel beams. The large number of rigidly connected joints provide the high lateral rigidity necessary to avoid a premium for the heights achieved. The Sears Tower in Chicago (1,450 ft. high) (442 meters) has 110 stories and is one of the three tallest buildings in the world utilizing the technology presented in this chapter.

If a plate structure element of a bridge has constant cross-section and its end support condition does not change transversely, the finite strip method has proven to be very efficient numerical structural analysis method, we are told by Cheung and Li in chapter 5. However, if the structure has any irregularities, e.g. a rectangular plate with openings, the finite strip method is no longer applicable on its own, and the finite element or boundary element method must be used. However, if these methods can be combined, with finite strips being used for the regular part of the plate and finite elements or boundary elements modelling the irregular part, then efficiency of the finite strip method and the universality of the latter methods are both utilized to their full advantage. These techniques and their combined utilization for the most effective analysis of complex bridge structures clearly show the significance and power of modern and constantly improving analysis technologies. Included are various numerical examples clearly demonstrating the effectiveness of the techniques presented.

In the final chapter of this book, by Albermani and Kitipornchai, we learn that thin-walled structural members are used widely in buildings and bridges because of their relatively light weight, ease of fabrication and construction, and availability in a variety of sectional shapes. These members, however, are highly susceptible to instability. Accurate determination of the buckling load, incorporating all possible buckling modes, local, distortional and flexural-torsional, is therefore very important. Because of the pervasiveness of structural elements and their fundamental nature, this chapter constitutes an essential element of this book.

This volume on structural dynamic systems techniques in buildings and bridges clearly reveals the effectiveness and essential significance of techniques available and, with further development, the essential role they will play in the future. The authors are all to be highly commended for their splendid contributions; these papers will provide a significant and unique reference source for students, research workers, practitioners, computer scientists and others on the international scene for years to come.

1 TECHNIQUES IN VIBRATION ANALYSIS OF STRUCTURAL STEEL FRAMES

GOMAN WAI-MING HO[1] and SIU-LAI CHAN[2]

[1] *Ove Arup & Partners (HK) Ltd., 56/F Hopewell Centre, 183 Queen's Road East, Hong Kong*
[2] *Department of Civil and Structural Engineering, Hong Kong Polytechnic University, Hung Hom, Hong Kong*

1.1. INTRODUCTION

Many high-rise buildings were constructed in the past decades due to rapid development, especially in South-East Asia. Financial considerations routinely call for the most economical structural solutions. Therefore, steel framed structures are receiving greater attentions because they offer shorter construction time, better seismic resistant, lighter self weight (cheaper foundation cost) etc. and are now becoming a more frequently adopted solution. However, with the trend toward constructing slender steel structures, the structural behavior is becoming more complex. Hence rigorous analysis methods were introduced to predict the behavior of various structures. The behavior of steel structures mainly depends on the material constitutive relationship of steel, geometry and connection details of the structure. This chapter discusses the effects of geometric nonlinearity and connection flexibility on the natural frequency of structural steel frames.

Extensive work has been carried out for static analysis of skeletal structures. Various numerical methods [Batoz and Dhatt (1979); Meek and Tan (1984); Yang and McGuire (1985); Chan (1988); Al-Bermani and

Kitpornchai (1988)] and formulations [Oran (1973); Chan (1989)] have also been developed for sophisticated pre- and post-buckling analysis of frames.

Apart from the material and geometric effects, the behavior of steel-structures is also dependent on joint stiffness. In conventional design and analysis of steel frames, the connections are commonly and conveniently assumed to be fully rigid or frictionless pinned. However, in the majority of engineering structures and buildings, the fabrication of a near fully rigid joint is expensive, impractical and economically unjustifiable in most cases. In reality, connections in steel frames are mostly semi-rigid and, consequently, the force and bending moment diagrams constructed under the rigid or pinned jointed assumption contain considerable error. The design based on these results will also lead to an inappropriate sizing of the members. In fact, joint flexibility has been considered by researchers for more than half a century ago [Batho and Rowan (1934); Rathbun (1936)]. However, these methods are too complex to use in practice and, as a consequence, the simplified assumption of either pinned or rigid joint is still used in practical design to date.

In various design codes for steel buildings such as the American Load Factor and Resistance Design and the British Standard BS5950, the effects of joint flexibility can be allowed for in an analysis. Research on the testing of various types of common connections as well as on the numerical techniques of conducting an analysis on flexibly connected steel frames were developed by Yu and Shanmugam (1986); Lui and Chen (1987a, 1987b); Nee and Haldar (1988); Pogg (1988); Goto and Chen (1987); Shi and Atluri (1989); Ho and Chan (1991, 1993). One of the simplest approaches for nonlinear analysis of frames with semi-rigid joints was developed by Lui and Chen (1987a) using the stability function. Ho and Chan (1993) used in conjunction with the finite element method and the Updated Lagrangian formulation for analysis of semi-rigid frames. Using these numerical procedures, the linear and nonlinear static analyses of frames with various types of flexible joints can be conducted accurately in both the pre- and post-buckling ranges.

The aforementioned analysis methods are mostly limited to static cases. It is recognized that a slender frame may have to be checked against dynamic instability in situations where resonance may occur and in seismic design. However, the research on these fields is very limited. One of the earliest theoretical contributions of dynamic response of semi-rigid frames was described by Lionberger and Weaver (1969) in 1963. A 10 storey frame was selected in their study and analyzed by a step-by-step method. In their study, the mass, moments of inertia were neglected and tributary masses were assumed to be concentrated at the framing levels. The connections were represented by rotational springs located between the central beam element and two rigid end bars with finite length. Sivakumaran (1988) studied the seismic response of a 10 storey and a 20 storey unbraced frames. Nonlinear dynamic analysis of these frames due to earthquake excitation was conducted. The characteristics of these frames were reported but the procedures of the numerical scheme was not detailed by Sivakumaran (1988).

Kawashima and Fujimoto (1984) first proposed the dynamic stiffness matrices for beam-column element composed of rotary springs at both ends. The internal degrees of freedom were eliminated by the transfer matrix method which finally yields the dynamic stiffness matrices with the same size as a rigidly connected structures with the same geometry. However, the formulation was limited to linear connection stiffness and the axial effects in the element were ignored.

Recently, Shi and Atluri (1989) presented a technique using the complementary strain energy for dynamic and nonlinear analysis for flexibly jointed frames. The linear vibration analysis and experiments on semi-rigid frames were also carried out by Kawashima and Fujimoto (1984). However, difficulties are encountered by using this element formulation due to the complicated procedure involved in including the rigid body motions in the consistent mass matrix. In other words, their techniques involved are not based on the widely used displacement based energy theories, leading to the difficulty for the engineer to directly extend his concepts of conventional analysis of framed structures to flexibly jointed frame analysis. Therefore, their method has not been applied to steel buildings with typical connection types.

In this chapter, a general procedure for deriving the element matrix of a beam-column element proposed by Chan and Ho (1994) is described. The method of incorporating a pair of connection springs to a conventional cubic Hermitian shape function is detailed. The special features about the proposed method are on the derivation and the use of a shape function with end springs to formulate the element matrices. By assigning a zero and a fully rigid spring stiffness, the displacement function will automatically convert to the linear and the cubic deflected shape with respectively zero and full nodal compatibility with the external rotations. Using this flexible shape function, consistent mass and stiffness matrices of a beam-column element can be obtained directly from the standard procedure for element formulation. The effects of joint flexibility are only included in the shape function whereas the other parts of the procedure follow exactly the same steps as in the development of a conventional cubic element. Therefore, the proposed technique for flexible joint analysis for frames is believed to be more readily accepted by the engineer for vibration checks of frames. In the present formulation for the consistent mass matrix, the displacement of a generic point is both a function of nodal point displacement and the deflection due to rotations about the convected axis passing through the two end nodes of an element. As these rotations are dependent on the connection stiffness and therefore cannot be pre-determined explicitly as a simple mathematical function, one must first incorporate the spring stiffness in the shape function as will be discussed in this chapter.

In the vibration check of slender frames, the stiffness cannot be assumed to be constant at all load levels, because of change of stiffness arising from the presence of initial axial stress. Further to this source of nonlinearity, the connection stiffness was measured experimentally to be a function of moment at a joint which in turn leads to a change of the overall stiffness of a structure when subject to a varying load. These two sources of nonlinearity will also be considered in this chapter.

1.2. BEHAVIOR OF CONNECTION

The function of connections is to transfer moments and forces from a structural member to other members and then to other parts of the structure or to the supports. They are also used to connect bracing to main members. A typical connection include connectors such as bolts, pins, rivets or welds and may also include additional plates or cleats.

Because the behavior of connections is a complex interaction of several components such as bolts and plates, the research and the development of connections in steel frames has been conducted for more than 70 years. Possibly the earliest experiment on the flexibility of riveted connections was carried out by Wilson and Moore (1917). During the 30's, studies on joint flexibility were actively conducted by Batho and Rowan (1934); Young and Jackson (1934) and Munse and his associates (1958, 1959, 1966).

A comprehensive review of the work on connections was given by Goverdhan (1934); Jones *et al.* (1983); Nethercot (1985) and Morris and Packer (1987) on some of the most commonly used connections. Summarizing this work, the torsional behavior of beam-to-column connections has not been examined in detail except for two studies carried out by Bennetts *et al.* (1978) and Grundy *et al.* (1983). It appears that torsional effects are not generally important in a framed structure and are therefore not included in this chapter. For the static loading case, a large number of tests were conducted to determine the moment-rotational relationship of connections. The tested connections can be basically grouped as the single cleat, double cleats, header plate, top and seat angle cleats, end plate, welded top plate and seat, T-stubs with and without web cleats connections and systems such as rivets, welds and bolts.

The cyclic behavior of moment-resisting connections was reported by Popov and Pinkey (1967a, 1967b, 1969) and sub-assemblages by Krawinker *et al.* (1971). The behavior of frames with realistic joints on reduced models was investigated by Carpenter and Lu (1973) and Clough and Tang (1975).

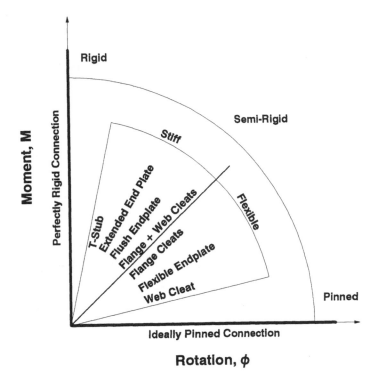

Figure 1. Range of stiffness of various connections.

As the details of connections used in practice could not be simulated accurately on a scaled model, Popov and Stephen (1970, 1976) therefore drew their discussion of cyclic moment connection behavior principally from the experiments of full scale models.

Test results have shown that most steel connections are neither rigid nor pinned, but semi-rigid. Pinned connections in practice normally possess some rotational stiffness, while connections classified as rigid often display some flexibility. It would therefore seem to be more correct to classify all steel frames under the heading of 'semi-rigid' jointed frames while recognizing 'simple' and 'continuous' construction as extreme cases. Figure 1 gives the range of stiffness for some commonly used connections.

The properties of connections are complex and uncertain in most cases. The effects of nonlinearity further complicate the problem. The practical sources of complexity includes geometric imperfections, lack of fit, residual stress due to welding, stress concentration and local secondary effects such as

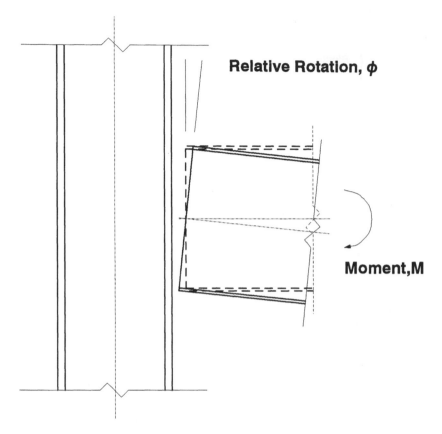

Figure 2. Rotational deformation of a connection.

panel zone deformation and strain hardening. In a typical bolt connection, the rotation is due to the bolt elongation, the column and beam flange deflection and the panel zone deformation etc. For most connections, the axial and shear deformations are usually small compared with the flexural deformation. For simplicity, only the rotational behavior of joints due to flexural action is considered here.

The primary in-plane flexural distortion of a steel beam-to-column connection can be simplified in Figure 2. The relative rotational deformation, ϕ, of a connection is caused by the applied beam-ended moment, M. This moment is transmitted from the beam to the column through the connection. In general, the in-plane stiffness of a connection is typically nonlinear and best described as a function of the moment, M and rotation, ϕ.

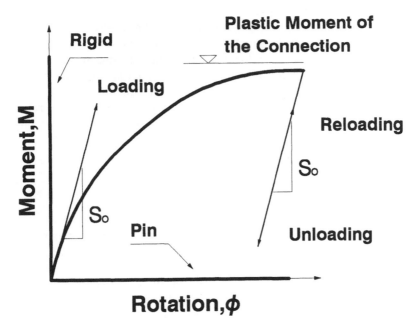

Figure 3. Theoretical moment-rotation curve of a typical connection.

The connection stiffness, S, decreases as the load increases and asymptotically approaches the plastic moment of the connection. If the loading applied to the structures causes reversal of moments and rotations, the initial stiffness of the connection is restored as shown in Figure 3. It is easily envisaged that unloading and reloading may take place at the joints, even though the frame is subject to a set of increasing loads. For example, the two ends of a beam will be under hogging moment from dead load, which will be reversed when the frame sways. This behavior becomes vital in the case of unbraced frames where wind loads may cause the joint to experience cyclic load.

The effect of joint unloading is more important if the frame is subject to dynamic forces. If the loading and unloading of a connection is repeated in a cyclic manner, its behavior can be represented by hysteretic loops. The consideration of cyclic joint behavior is important in time history analysis of steel frames, but is outside the scope of this chapter and discussed in a companion chapter of this series "Techniques in transient analysis of semi-rigid steel frames".

To study the influence of joint flexibility in frame structures, the joint behavior must first be incorporated into an analysis. Using the finite element method, a structure can be idealized as an assemblage of plates and shells. The stiffness for various components such as bolts and beam-columns can then be assembled to obtain the structural stiffness for analysis. This rigorous approach, however, consumes too extensive computer time and data input and output efforts. Consequently, it is not feasible for practical design. An alternative approach is to employ only the beam-column and connection element to model a frame. The successful use of this latter method relies heavily on the available M-ϕ data for various joints used in connecting members in a frame.

1.3. MODELLING OF CONNECTIONS

The literatures mentioned in the previous section focus on the behavior of connections based on experimental results. The characteristics or the strength of connections with different combination of components was determined in tests. The analytical formulation was first suggested by Lothers (1951, 1960) for computing the initial stiffness of semi-rigid connections. The initial connections stiffness predicted by this method was in good agreement with the experimental results by Rathbun (1936). This idea was further developed by Yu (1953) and Yu (1955) on top and bottom seated angle connections. Huang (1958) and Kenndy and Hafez (1984) also produced their results on header plate connections.

To predict the nonlinear connection behavior, Wales and Rossow (1983) proposed a model for double web cleat connections with nonlinear axial springs. Afterward, this model was modified by Chmielowiec and Richard (1987) for all types of cleated connections. Because these approaches were developed on the assumed connections mechanism, yield line theory or nonlinear axial springs, these non-linear models can only be applied to web and/or cleated angle connections.

Using the finite element method, the behavior of various types of connections could be modelled and analyzed. Bose et al. (1972) first used a finite element model to predict the behavior of typical connections. Patel and Chen (1984) employed the plane stress element in modelling fully welded connections. The residual stresses in the connections were included in their studies using the computer package NONSAP developed by Bathe and Wilson (1973). Recently, Krishnamurthy et al. (1979), and Kukreti et al. (1987) proposed similar finite element models for extended end plate and flush end plate connections. Because the discussed methods are based on an assumed connection mechanism or on a finite element model, they are grouped into that class of analytical model where experiments are not required to determine the stiffness of connections.

Normally analytical models can only provide an approximate prediction of actual behavior of connections due to uncertainties in the joints. Based on numerous investigations into the behavior of semi-rigid connections, a data bank for M-ϕ curves has been generated. Therefore, it is more advantageous to curve-fit the experimental data by a mathematical function for the purpose of tracing the M-ϕ curve. A very accurate representation of the connection M-ϕ curves was proposed by Jones et al. (1982) using the cubic B-spline technique. However, the number of parameters of this model is enormous and may involve excessive computational effort. Lui (1983) suggested that the connection behavior should be modelled by a series of exponential functions which were later refined to include the effects of strain-hardening by Kishi and Chen (1986). Because the B-spline and the modified exponential models can only be used to curve-fit the available data set, they are only mathematical functions without physical meaning, properties or characteristics.

Although extensive work has been conducted on the studies of the M-ϕ relationship for various types of connections, the scope of these tests may not cover all types of connections and thus causes inconvenience to the designers. Sommer (1969) first fitted the M-ϕ data points of the header plate connections in the form of a non-dimensional polynomial function. This is the so-called standardized M-ϕ function from which the behavior of similar types of connections can be predicted. This approach was later extended by Frye and Morris (1975) to seven different types of joints. Because the polynomial functions adopted by Sommer (1969) and Frye and Morris (1975) are to measure the peak and trough of the available M-ϕ data, the functions may yield a negative connection stiffness which is generally unacceptable in an analysis. To eliminate the occurrence of the negative connection stiffness, Ang and Morris (1984) modified the Frye-Morris model (1975) in combination with the Ramberg-Osgood function (1943). The concept of standardized connection models was recently refined by Attiogbe and Morris (1991) in a least-squares fitting procedure. The connections behavior can also be expressed in a power form suggested by Kishi and Chen (1987) and Colson and Louveau (1968) on the basis of the function proposed by Richard and Abbott (1975). Because these models have the properties of predicting the connection behavior with parameters based on the experimental results, they are grouped into the class of mixed model which have characteristic lying between the analytical and the mathematical models.

Generally speaking, the joint behavior can be simplified to a set of moment versus rotation, M-ϕ, relationships. Mathematically, these relations can be represented in a general form as,

$$M = f(\phi) \tag{1}$$

or, inversely,

$$\phi = g(M) \tag{2}$$

in which M is the applied moment; and ϕ is the slip angle equal to the difference between the angles at the two ends of a connection.

Classification of the moment versus rotation relations for joints in accordance with different mathematical functions such as linear, nonlinear power or series may be inappropriate in terms of the physical characteristics of a joint model. In contrast to this grouping criterion, it appears to be more germane to divide connection models into those based on analytical, mathematical or hybrid methods of derivation. The following sub-section describes the features of these models. For clarity, a summary of the available connection models for various type of connections is tabulated in Table 1.

1.3.1. Analytical Models

Analytical models are derived to predict the joint stiffness on the basis of the interaction between geometrical properties of the components. Assuming a connection mechanism, some methods such as the finite element method could be used to predict the behavior of the connection. These models are therefore commonly used to conduct parametric studies on the effect of various geometric and force variables related to the components of the connections. Practical ranges of these variables are then analyzed to produce data for actual analysis. A simplified procedure to determine the connection stiffness is then derived and incorporated into the an analysis software. However, due to the extensive number of parameters involved in this type of analytical approach, it is not widely used.

1.3.2. Mathematical Models

Currently, the most commonly used approach to describe the M-ϕ relationship is to curve-fit the experimental data with simple expressions. Numerous experiments on connections have been performed in the past and, using the curve-fitting technique, a rather large body of M-ϕ data is now available.

Mathematical models have been developed to curve-fit the available data. The Exponential model by Lui (1985) and its modified version by Kishi and Chen (1986) were developed on the basis of the least-squares method. These procedures can trace the experimental data accurately but the functions are very sensitive to the scattered experimental data. Four types of common connections with their M-ϕ curves are shown in Figure 4. Namely, they are the single web (A), the top and seated angle (B), the flush end plate (C) and the

Table 1. Connection Models for some Typical Connections

	No. of Coefficients in a General Expression[†]	M-φ Curve	Type of Connections								
			Single Web Angle	Double Web Angle	Header Plate	Top and Seat Angle	Top and Seat Angle with Double Web Angle	End Plate	End Plate with Column Stiffness	T-Stub	Strap Angle
Analytical Model											
Lothers (1951)	5	Linear / Bi-linear	•			•	•				
Johnson-Law (1981)	5	Linear / Bi-linear			•					•	
Krishnamurthy et al. (1979)	N/A	Nonlinear		•							
Wales-Rossow (1983)	N/A	Nonlinear		•							
Kenndy-Hafez (1984)	N/A	Nonlinear						•			
Kukreti et al. (1987)	N/A	Nonlinear						•			
Azizinamini et al. (1987)	N/A	Linear / Bi-linear				•	•				
Mixed Model											
Frye-Morris (1975)	4	Nonlinear	•	•	•	•	•	•	•		
Ang-Morris(1984)	4	Nonlinear	•	•	•	•					•
Yee-Melchers (1986)	4	Nonlinear	•	•							
Kishi-Chen (1987)	3	Nonlinear	•	•		•	•	•	•		
Wu-Chen(1990)	3	Nonlinear				•	•			•	
Attiogbe-Morris (1991)	4	Nonlinear	•	•	•	•	•	•	•	•	•

Notes:

The models which the number of coefficients are marked with N/A can not be represented by a single general equation.

Mathematical models for connections were not shown in the table because they can be used for all types of connections if the experimental data were available.

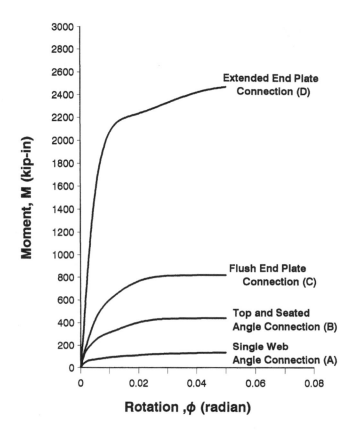

Figure 4. $M - \phi$ curves of some common connections.

extended end plate connections (D). The nonlinear behavior of these joints was measured experimentally and modelled by the exponential function given by Lui and Chen (1988). Theoretically, the stiffness of connections should decrease with an increase of moment. However, the lack of fit in a joint may result in an increase of stiffness at a very low load level. This effect may be eliminated by pre-loading in the experiment. From the design point of view, this effect can be ignored. In spite of this observation, the mathematical curve by Lui and Chen (1988) possesses an increase in stiffness at zero load, as shown in Figure 5 and Figure 6. Because of this, the buckling load of a frame may be increased when the moment at joint is small. To this end, the commonly adopted model of replacing a set of loads by two equivalent nodal point loads at connections may not create significant moment at the

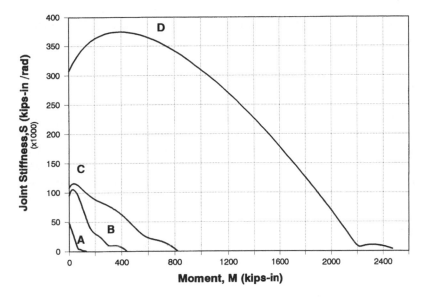

Figure 5. Joint stiffness-moment curves of some common connections (Modelled by modified exponential functions).

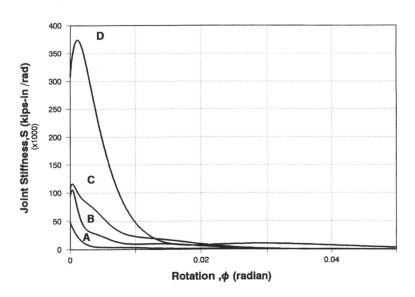

Figure 6. Joint stiffness-rotation curves of some common connections (Modelled by modified exponential functions).

connections, whilst the actual load would generate a significant moment at connections. This contrast in moments at joints may lead to an overestimated bifurcation load for a frame.

1.3.3. Mixed Models

Mixed models are the combinations of the analytical and the mathematical models based on experimental results. Most of the available connection models are under this category. Examples for this group of models are those models proposed by Frye and Morris (1975); Ang and Morris (1984); Kishi and Chen (1987); Yee and Melchers (1986) and Wu and Chen (1990). These models contain the so-called standardized functions or at least one 'curve-fitting' parameter which must be obtained from experiments. Hence the mixed models are functions of the geometry and the components of the connections. Because the general expression of a mixed model contains very few parameters, each model can only be used for one or two particular types of connections. These models in general can be used to calculate the initial stiffness of particular types of connections, which is impossible for mathematical models, and predict the nonlinear behavior of connections.

1.4. DERIVATION OF CONNECTION ELEMENT MATRIX

The stiffness of the joint, S, is the slope of the moment versus rotation (M-ϕ) curve at relative angle, ϕ, and expressed as:

$$S = \frac{dM}{d\phi} \tag{3}$$

In the case of moment reversal occurring at a joint (i.e. $M \cdot dM < 0$), the stiffness of the connection is assumed to be equal to the initial joint stiffness, S_0. Thus:

$$S_\phi = S_0 = \frac{dM}{d\phi}\bigg|_{\phi=0} \qquad \text{if } M \cdot dM < 0 \tag{4}$$

In the derivation of the spring stiffness matrix, the three basic governing conditions are made use of. These are the compatibility, the equilibrium and the constitutive relations for an element. Considering a spring with stiffness S, two end rotations, $_e\theta$ and $_i\theta$, and moments, $_em$ and $_im$, as shown in Figure 7, the equilibrium condition requires that:

$$_em +_i m = 0 \tag{5}$$

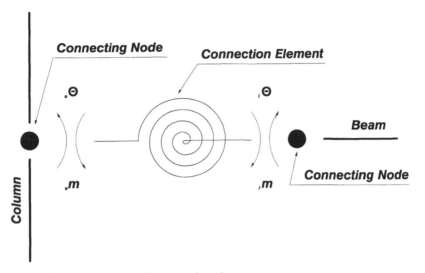

Figure 7. Nomenclature of a connection element.

Assuming that the connection element is dimensionless and ϕ is the slip angle between the two end rotations and following the compatibility condition, we have:

$$\phi = {}_e\theta - {}_i\theta \tag{6}$$

The constitutive relationship can finally be stated as:

$$S = \frac{{}_em}{\phi} = \frac{{}_im}{\phi} \tag{7}$$

Combining Equations (5) to (7), we have:

$$S = \frac{{}_em}{{}_e\theta - {}_i\theta} = \frac{{}_im}{{}_e\theta - {}_i\theta} \tag{8}$$

or, in matrix form,

$$\begin{bmatrix} S & -S \\ -S & S \end{bmatrix} \begin{bmatrix} {}_e\theta \\ {}_i\theta \end{bmatrix} = \begin{bmatrix} {}_em \\ {}_im \end{bmatrix} \tag{9}$$

Equation (9) is the equilibrium equation of the connection element. Therefore, the stiffness matrix of the connection element is given by :

$$\begin{bmatrix} S & -S \\ -S & S \end{bmatrix} \tag{10}$$

1.5. DERIVATION OF FLEXIBLE SHAPE FUNCTION FOR BEAM-COLUMN

In the displacement based finite element formulation, a mathematical function for the deflection is first assumed. For beam-column type of one-dimensional element, the cubic Hermitian function is most commonly used, although other shape functions have been proposed So and Chan (1991). In this chapter, the conventional cubic Hermitian function is employed for its simplicity and publicity. However, it must be emphasized that other functions can also be used in the formulation of element with end springs by the present procedure.

The lateral deflections, v, on the y-direction and x, along the centre line of a straight element can be written as:

$$v = a_0 + a_1 x + a_2 x^2 + a_3 x^3 \tag{11}$$

Assuming the coordinate axis to pass through the two nodes of the element, the deflection, v_1 and v_2, can be assumed to vanish and the deflection can be expressed in terms of the nodal rotations through the following conditions:

$$x = 0, \quad v = 0, \quad \frac{dv}{dx} = {}_i\theta_1^z \tag{12}$$

$$x = L, \quad v = 0, \quad \frac{dv}{dx} = {}_i\theta_2^z \tag{13}$$

in which ${}_i\theta_1^z$ and ${}_i\theta_2^z$ are the angles between the tangents to the curve at the two ends and the chord connecting the two nodes of the element as shown in Figure 8.

Solving Equations (11) to (13), the lateral deflection, v, can be written as,

$$v = \rho_1^2 \rho_2 L_i \theta_1^z - \rho_1 \rho_2^2 L_i \theta_2^z \tag{14}$$

in which ρ_1 and ρ_2 are dimensionless parameters given by,

$$\rho_1 = \left(1 - \frac{x}{L}\right) \text{ and } \rho_2 = \frac{x}{L} \tag{15}$$

From the consideration of geometrical compatibility, the displacement of an arbitrary point along the deflected curve is equal to the sum of the deflections due to the end nodal rotations and the translational nodal displacements (see Figure 8) as,

$$v = \{\rho_1^2 \rho_2 L - \rho_1 \rho_2^2 L\} \{{}_i\theta_1^z \ {}_i\theta_2^z\}^T + \rho_1 v_1 + \rho_2 v_2 \tag{16}$$

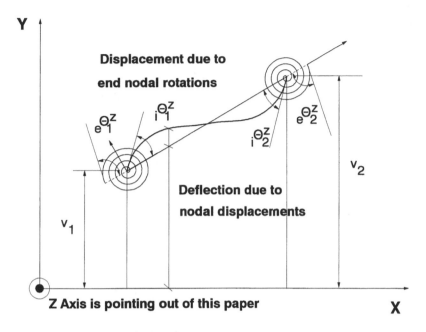

Figure 8. Deflections and rotaions of an element (Similar for X-Z planr).

Note that an infinitesimal inclination of the displaced axis to the undeformed axis is assumed in Equation (16). Hence the deflection in Equation (14) can be directly added to the displacement due to nodal deflections.

The rotations at the two external nodes of an element, denoted as $_e\theta_1^z$ and $_e\theta_2^z$, are not generally equal to the element internal nodal rotations, $_i\theta_1^z$ and $_i\theta_2^z$ (see Figure 8). This is because of the geometrical incompatibility arisen from joint flexibility. In other words, the slip angle, ϕ, exists between the two ends of a connection or joint.

The element matrix for the complete system composed of two connection springs and the beam-column element can be written as

$$
\begin{bmatrix}
S_1^z & S_1^z & 0 & 0 \\
-S_1^z & S_1^z + k_{11}^z & k_{12}^z & 0 \\
0 & k_{21}^z & k_{22}^z + S_2^z & -S_2^z \\
0 & 0 & -S_2^z & S_2^z
\end{bmatrix}
\begin{bmatrix}
_e\theta_1^z \\
_i\theta_1^z \\
_i\theta_2^z \\
_e\theta_2^z
\end{bmatrix}
=
\begin{bmatrix}
_em_1^z \\
_im_1^z \\
_im_2^z \\
_em_2^z
\end{bmatrix}
\tag{17}
$$

in which k_{ij}^z are the coefficients of bending stiffness, S_1^z and S_2^z are the joint stiffness; $_e\theta_1^z$, $_e\theta_2^z$, $_i\theta_1^z$ and $_i\theta_2^z$ are respectively the external and internal rotations at nodes 1 and 2 shown in Figure 8. k_{ij}^z are the coefficients of bending stiffness given by,

$$k_{ij}^Z = \frac{EI}{L}\begin{bmatrix} 4 & 2 \\ 2 & 4 \end{bmatrix} \tag{18}$$

Considering only the two equations for internal rotations, we have:

$$\begin{bmatrix} k_{11}^z + S_1^z & k_{12}^z \\ k_{21}^z & k_{22}^z + S_2^z \end{bmatrix}\begin{bmatrix} {}_i\theta_1^z \\ {}_i\theta_2^z \end{bmatrix} + \begin{bmatrix} -S_1^z & 0 \\ 0 & -S_2^z \end{bmatrix}\begin{bmatrix} {}_e\theta_1^z \\ {}_e\theta_2^z \end{bmatrix} = \begin{bmatrix} {}_im_1^z \\ {}_im_2^z \end{bmatrix} \tag{19}$$

For simplicity, the loads are assumed to apply at the global nodes and therefore, Equation (19) can be simplified as:

$$\begin{bmatrix} {}_i\theta_1^z \\ {}_i\theta_2^z \end{bmatrix} = \begin{bmatrix} k_{11}^z + S_1^z & k_{12}^z \\ k_{21}^z & k_{22}^z + S_2^z \end{bmatrix}^{-1}\begin{bmatrix} S_1^z \cdot {}_e\theta_1^z \\ S_2^z \cdot {}_e\theta_2^z \end{bmatrix} \tag{20}$$

Substituting Equation (20) into Equation (16) , we have:

$$v = (\rho_1^2\rho_2L - \rho_1\rho_2^2L)\begin{bmatrix} k_{11}^z + S_1^z & k_{12}^z \\ k_{21}^z & k_{22}^z + S_2^z \end{bmatrix}^{-1}\begin{bmatrix} S_1^z \cdot {}_e\theta_1^z \\ S_2^z \cdot {}_e\theta_2^z \end{bmatrix} + \rho_1v_1 + \rho_2v_2 \tag{21}$$

In Equation (21), the variables are in the element local co-ordinate system that the effect of joint flexibility has been allowed for in the assumed deflections of the element. For example, when the connection stiffness is infinite, the deflected shape is in the form of a Hermitian cubic function dependent on the nodal variables. If the connection stiffness vanishes, the deflected shape becomes a straight line as in the case of a pin-jointed truss member. The condition of a semi-rigid connection will lie between these two extremes.

The rotations, ${}_e\theta_1^z$ and ${}_e\theta_2^z$, about the convected axis passing through the two ended nodes can further be expressed by the global variables through the consideration of geometrical compatibility. This is essential to formulate the element matrix in the global co-ordinate system for assembling the global structural stiffness matrix. Thus, the lateral deflection, v, can finally be written as,

$$v = (\rho_1^2\rho_2L - \rho_1\rho_2^2L)$$

$$\begin{bmatrix} k_{11}^z + S_1^z & k_{12}^z \\ k_{21}^z & k_{22}^z + S_2^z \end{bmatrix}^{-1}\begin{bmatrix} S_1^z & 0 \\ 0 & S_2^z \end{bmatrix}\begin{bmatrix} \frac{1}{L} & 1 & -\frac{1}{L} & 0 \\ \frac{1}{L} & 0 & -\frac{1}{L} & 1 \end{bmatrix}\begin{bmatrix} v_1 \\ {}_e\theta_1^z \\ v_2 \\ {}_e\theta_2^z \end{bmatrix} \tag{22}$$

$$+ \rho_1v_1 + \rho_2v_2$$

The expression in (22) represents the relationship between the lateral deflection, v, the connection stiffness, S_1^z and S_2^z, and the nodal variables. This can be used directly to formulate the linear, the geometric stiffness and the mass matrices. The procedure for deriving a displacement based element can be made through the application of the principle of total potential energy [e.g. Meek and Tan (1984) and Chan and Kitipornchai (1986)] or the virtual work principle [e.g. Chan (1989) and Bathe and Bolourchi (1979)]. Because of this, it is believed that the present approach is simply a direct extension of the rigid frame analysis to cover a more general case of semi-rigid connection. Consequently, it is envisaged that the present method will be easily accepted by analysts and engineers.

1.6. DERIVATION OF STIFFNESS MATRICES

In this section, the variational principle will be adopted to develop the stiffness matrices using the proposed flexible shape function. The variational principle was first developed some 200 years ago [see Oden and Reddy (1976)]. Its application to elasticity and plasticity problems has been described in a well-known text by Washizu (1982). The variational method and energy theorems have been widely used by researchers in the finite element method, including Zienkiewicz (1989); Bathe (1982) and Reddy (1984, 1986) and many others. In this chapter, the theorem of minimum potential energy is adopted for the formulation of the stiffness matrices.

The total potential energy π of the system is expressed as,

$$\prod = U - W \tag{23}$$

in which U is the strain energy of a continuum and W is the external work done.

The external work done, W, is defined as the product of the applied forces, F_i and the corresponding displacements, u_i. In matrix form, this relationship can be written as,

$$W = [u]^T [F] \tag{24}$$

in which $[F]$ is the equivalent nodal applied forces for all the external loads including surface traction and body forces and $[u]$ is the nodal displacements.

The strain energy, U, can be written in terms of the stress tensor $[\sigma]$ and strain tensor $[\epsilon]$ as,

$$U = \frac{1}{2} \int_{\text{vol}} [\sigma][\epsilon] d(\text{vol})$$
$$= \frac{1}{2} \int_{\text{vol}} [\epsilon]^T [D][\epsilon] d(vol) \tag{25}$$

where $[D]$ is the matrix of Hookean's material constants relating the material properties and vol is the volume of the continuum.

The strain tensor $[\epsilon]$ may be decomposed into two parts as,

$$[\epsilon] = [\epsilon]^L + [\epsilon]^{NL} \tag{26}$$

where the superscripts L and NL represent the linear and nonlinear strain components respectively.

Substituting Equation (26) into Equation (25) gives,

$$U = \frac{1}{2} \int_{vol} [\epsilon^L + \epsilon^{NL}]^T [D][\epsilon^L + \epsilon^{NL}] d(\text{vol}) \tag{27}$$

Expanding Equation (27) gives,

$$U = \frac{1}{2} \int_{vol} \left([\epsilon^L]^T [D][\epsilon]^L + 2[\epsilon^L]^T [D][\epsilon^{NL}] + [\epsilon^{NL}]^T [D][\epsilon^{NL}] \right) d(\text{vol}) \tag{28}$$

Neglecting the higher order nonlinear strain product, the strain energy U can be expressed in matrix form as,

$$U = \frac{1}{2} \int_{vol} \left([\epsilon^L]^T [D][\epsilon^L] + 2[\sigma^L]^T [\epsilon^{NL}] \right) d(\text{vol}) \tag{29}$$

or

$$U = U^L + U^{NL} \tag{30}$$

in which U^L and U^{NL} are the linear and nonlinear parts of U. The stress and strain tensor $[\sigma^L]$ and $[\epsilon^L]$ corresponding to the linear stress and strain respectively can be related as,

$$[\sigma^L] = [D][\epsilon^L] \tag{31}$$

According to the principle of minimum potential energy and assuming the system is conservative and the load $[F]$ is independent of $[u]$, the variation of the functional with admissible or state variables u is ,

$$\delta \prod = \frac{\partial \prod}{\partial u}$$

$$= \frac{\partial U^L}{\partial u} + \frac{\partial U^{NL}}{\partial u} - [F] = 0 \tag{32}$$

The second variation of the functional is,

$$\delta^2 \prod = [K_L] + [K_G] \tag{33}$$

in which $[K_L]$ and $[K_G]$ are respectively the linear and geometric stiffness matrices given by the following expressions,

$$[K_{Lij}] = \frac{\partial^2 U^L}{\partial u_i \partial u_j} \tag{34}$$

and

$$[K_{Gij}] = \frac{\partial^2 U^{NL}}{\partial u_i \partial u_j} \tag{35}$$

To obtain the solution for the finite element method using the variational principle or the energy method, approximation function(s) must first be assumed as,

$$u_e \approx \tilde{u}_i = \sum_{i=1}^{n} \psi_i \bar{u}_i \tag{36}$$

where n is the total number of degrees of freedom of an element, u_e is a generic point approximated by $\tilde{u}_i$. $\bar{u}_i$ represents the element nodal displacements, u_i, v_i, w_i and ϕ_i, which are equal to the state variables. ψ_i is the shape, the approximated or the interpolation functions. Each of the term ψ_i defines the variation of element displacement, u_e, with respect to the corresponding degree of freedom. The function, ψ_i, will be equal to unity at i degree of freedom and zero at other nodes.

In the present study, a linear interpolation function is assumed for the axial lengthening and the twist whilst the flexible shape function, which was developed in the previous section and based on a cubic Hermite polynomial function, will be used for interpolating the lateral displacement.

Once the shape functions, ψ_i are sought, the strains of an element, $[\epsilon_e]$ can be obtained from displacement as,

$$[\epsilon_e] = \frac{\partial [u]}{\partial x} \tag{37}$$

Symbolically, it can be written as,

$$[\epsilon_e] = [B][u] \tag{38}$$

and

$$[B] = \frac{\partial \psi_i}{\partial x} \qquad \text{for } i = 1, \ldots n \tag{39}$$

where $[B]$ is the strain-displacement matrix and $[u]$ is the nodal displacement vector.

Substituting Equation (38) into Equation (25), the governing equation can be related to the element nodal degrees of freedom, external nodal forces and element stiffness. The $[B]$ matrix can further be decomposed into the linear and the nonlinear parts symbolically as,

$$[B] = [B^L] + [B^{NL}] \tag{40}$$

where $[B^L]$ and $[B^{NL}]$ are the linear and nonlinear strain-displacement matrices. Substituting Equations (38) and (40) into Equation (30) gives,

$$
\begin{aligned}
U = U^L + U^{NL} \\
= \frac{1}{2} \int_{\text{vol}} [u]^T [B]^T [D][B][u] d(\text{vol}) \\
= \frac{1}{2} \int_{\text{vol}} [u]^T [B^L][D][B^L][u] d(\text{vol}) \\
+ \frac{1}{2} \int_{\text{vol}} [u]^T [B^L][D][B^{NL}][u] d(\text{vol})
\end{aligned}
\tag{41}
$$

$$+ \text{ higher order terms}$$

Hence, the second variation of U^L or the linear stiffness matrix is given by,

$$
\begin{aligned}
[K_L] &= \frac{\partial^2 U^L}{\partial u_i \partial u_j} \\
&= \frac{\partial^2}{\partial u_i \partial u_j} \left[\frac{1}{2} \int_{\text{vol}} [u]^T [B^L]^T [D][B^L][u] d(vol) \right] \\
&= \int_{\text{vol}} [B^L][D][B^L] d(\text{vol}) \\
&= \int_{\text{vol}} [\sigma][B^L] d(\text{vol})
\end{aligned}
\tag{42}
$$

and the geometric stiffness matrix, $[K_G]$ is as follows,

$$
\begin{aligned}
[K_G] &= \frac{\partial^2 U^{NL}}{\partial u_i \partial u_j} \\
&= \frac{\partial^2}{\partial u_i \partial u_j} \left[\frac{1}{2} \int_{\text{vol}} [u]^T [B^L]^T [D][B^{NL}][u] d(vol) \right] \\
&= \int_{\text{vol}} [B^L][D][B^{NL}] d(\text{vol}) \\
&= \int_{\text{vol}} [\sigma][B^{NL}] d(\text{vol})
\end{aligned}
\tag{43}
$$

For beam-column element,

$$D = E$$

$$\epsilon = -\frac{d^2v}{dx^2} \tag{44}$$

$$\sigma = -EI\frac{d^2v}{dx^2}$$

The linear stiffness matrix $[k_l]$ for a beam-column element can be simply expressed as,

$$[k_l] = \int_L \left[\frac{d^2\psi}{dx^2}\right]^T EI \left[\frac{d^2\psi}{dx^2}\right] dx \tag{45}$$

where EI is the flexural rigidity and L is the length of the element. The coefficients of the elastic stiffness matrix, $[k_l]$ are listed in Appendix 1.

Similarly, the geometric stiffness matrix for a beam-column element, $[k_g]$ is written by,

$$[k_g] = \int_L \left[\frac{d\psi}{dx}\right]^T \frac{P}{2} \left[\frac{d\psi}{dx}\right] dx \tag{46}$$

where P is the axial force of the member and the coefficients of the geometric stiffness are listed in Appendix 2.

1.7. DERIVATION OF CONSISTENT MASS MATRIX

For a dynamic system, inertia forces must be taken into account and the kinetic energy of the continuum is given by ,

$$K.E. = \frac{m\dot{u}^2}{2} \tag{47}$$

where m is the mass of the continuum and u is the velocity of the continuum.

Equation (47) can also be written in matrix form as,

$$K.E. = \frac{1}{2} \int_{vol} [\dot{u}]^T \rho [\dot{u}] dvol \tag{48}$$

where ρ is the density of the material, vol is the volume of the continuum and $[u]$ is the vector of nodal velocity. Consider the case of free vibration where the nodal displacement is time dependent,

$$[u(t)] = Ae^{i\omega t} \tag{49}$$

where A is a constant, ω is the natural frequency of vibrations; $i = \sqrt{-1}$ and $t =$ time.

Hence, the first derivative of Equation (49) gives,

$$[\dot{u}] = i \cdot \omega \cdot Ae^{i\omega t}$$
$$= i \cdot \omega \cdot [u(t)]$$

(50)

As

$$[u(t)] = [\psi][u]$$

(51)

where $[\psi]$ is the matrix of shape function, Equation (50) becomes

$$[\dot{u}] = i \cdot \omega \cdot [\psi][u]$$

(52)

Hence, the expression of the kinetic energy in matrix form is:

$$K.E. = -\frac{1}{2} \int_{\text{vol}} [u]^T [\psi]^T \rho [\psi][u] d\text{vol}$$

(53)

For a prismatic section $\rho\, d$vol can be replaced by:

$$\rho\, d\text{vol} = \overline{m} dx$$

(54)

where $\overline{m}$ is the mass per unit length of the member. Because the shape function was also represented by the assumed displacement function, v. The consistent mass matrix is therefore written as:

$$[m] = \int_L [\psi]^T \overline{m} [\psi] dx$$

(55)

where $[m]$ is the consistent mass matrix for the beam-column element and the coefficients are listed in Appendix 3.

1.8. FREE VIBRATION EQUATIONS

In vibration analysis, the displacement is assumed to be a function of both the distance along the centroidal axis of the element, x, and of time, t. For clarity, the procedures are demonstrated on the $x - y$ plane. Thus, the modified displacement function taking into account the time effect is assumed as,

$$
\begin{aligned}
v = \sin(\omega t - \alpha) &\left\{ \{\rho_1^2 \rho_2 L - \rho_1 \rho_2^2 L\} \begin{bmatrix} k_{11}^z + S_1^z & k_{12}^z \\ k_{12}^z & k_{22}^z + S_2^z \end{bmatrix}^{-1} \begin{bmatrix} S_1^z & 0 \\ 0 & S_2^z \end{bmatrix} \right. \\
&\left. \begin{bmatrix} \frac{1}{L} & 1 & -\frac{1}{L} & 0 \\ \frac{1}{L} & 0 & -\frac{1}{L} & 1 \end{bmatrix} \begin{bmatrix} v_1 \\ {}_e\theta_1^z \\ v_2 \\ {}_e\theta_2^z \end{bmatrix} + \rho_1 v_1 + \rho_2 v_2 \right\}
\end{aligned}
$$

(56)

in which t is the time and α is the phase angle.

The free vibration equilibrium equation of a skeletal structure can be written as:

$$[M][\ddot{v}] + [K_L][v] = [0] \tag{57}$$

in which is $\ddot{v}$ the acceleration; $[M]$ and $[K_L]$ are respectively the mass and the linear stiffness matrices for the structure. Hence:

$$[M] = \sum_{i}^{n} [m]$$

$$[K_L] = \sum_{i}^{n} [k_l] \tag{58}$$

The frequency of the system can then be obtained via the non trivial condition for the equilibrium Equation (57) as:

$$|-\omega^2[M] + [K_L]| = 0 \tag{59}$$

1.9. NONLINEAR FREE VIBRATION ANALYSIS

Strictly speaking, the stiffness of a framed structure is not constant but dependent on the load. The axial effects can be included in the instantaneous or incremental equilibrium equation via the adoption of the geometric stiffness matrix. The natural frequency, ω, determined on the basis of this set of equations can be written as:

$$|-\omega^2[M] + [K_L] + \lambda[K_G]| = 0 \tag{60}$$

in which λ is a load intensity parameter.

Similar to the semi-bifurcation approach as mentioned by Ho and Chan (1991), the connection stiffness may be nonlinear and dependent on the moment at joints. Therefore an iterative procedure will be required to calculate the stiffness at a particular load level. Mathematically, Equation (60) is replaced by:

$$|-\omega^2[M] + [K_L(\lambda)] + [K_G]| = 0 \tag{61}$$

in which $[K_L(\lambda)]$ represents the stiffness matrix being a function of moments at joints, and therefore indirectly dependent on the applied load.

1.10. NUMERICAL PROCEDURE FOR VIBRATION ANALYSIS

In the eigenvalue analysis of frames taking into account the nonlinear effects due to the initial stress and joint stiffness, a preliminary linear analysis is first conducted. At this stage, the moments at the elements are determined on the basis of the initial joint stiffness. These moments are then used to update the connection stiffness which, in general, is a nonlinear function of the applied moment. The resistance, $[R]$, of the structure against the applied loads is then computed according to these updated or corrected joint stiffness as,

$$[R] = \sum_{j=1}^{n} \int_{\text{vol}} [B]^T [\sigma(\lambda, v)] d(\text{vol}) \qquad (62)$$

in which $[B]$ is the strain versus displacement transformation matrix, $[\sigma(\lambda, v)]$ is the internal, true or the Cauchy stress of the element and is calculated from the material properties, the joint rotations and the updated joint stiffness.

This iterative procedure is repeated until the error of the analysis, measured by the difference between the Euclidean norms of the resistant forces, is less than a certain tolerance of, say, 0.1% of the applied force. After the satisfaction of the equilibrium condition, the analysis will be directed to another subroutine for the solution of eigenvalue equations and for the computation of the natural frequency, ω.

In this eigenvalue solution subroutine, the change of the natural frequencies in two consecutive iterations is determined and if it is found to be very small, the convergence to the semi-definite condition for the instantaneous stiffness matrix is assumed to have been satisfied. Otherwise, the next estimated natural frequency is predicted by the implicit polynomial scheme and the iterative procedure is continued until convergence.

1.11. NUMERICAL EXAMPLES

To validate the proposed theory, a number of linear and nonlinear problems were selected and solved by the computer program developed on the basis of the suggested method. Special emphasis was placed on the analysis of frames with semi-rigid connections in which the commonly assumed pinned and rigid joints are only the extreme cases.

1.11.1. Convergence Test of Pinned, Fixed and Flexibly Jointed Beams

The beams with ends pinned, fixed and flexible spring of stiffness equal to EI/L as shown in Figure 9 are analyzed and the natural frequencies are determined. The error for the predicted natural frequency by using 2, 4 and 8 elements is also shown in the same figure. The analytical solutions for

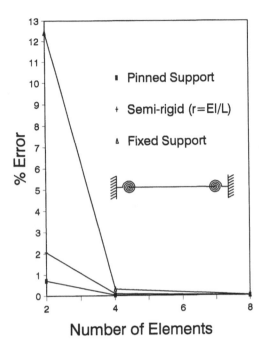

Figure 9. Convergence test for numerical frequency of beam.

the pinned and fixed end cases are $\pi_2\sqrt{(EI/\overline{m}L^4)}$ and $22.3733\sqrt{(EI/\overline{m}L^4)}$ respectively in which $\overline{m}$ is the mass per unit length. As expected, this error for the computed frequency of the flexibly jointed beam lies between the extreme cases of pinned and rigid joints, because of the error of fitting a cubic curve onto the deflected shape of the beam. This phenomenon was also noted by So and Chan (1991), in the bifurcation analysis of columns by the third and fourth order elements.

1.11.2. Natural Frequencies of Beams with Varying Connection Stiffness

In this example, the natural frequency for a beam-column member with two ends flexibly connected is studied. The results of the change in natural frequency due to the variation of joint stiffness is shown in Figure 10. The frequency of the beam is represented by the square root of its flexural rigidity, mass per unit length and the quadruple of length. When the spring stiffness

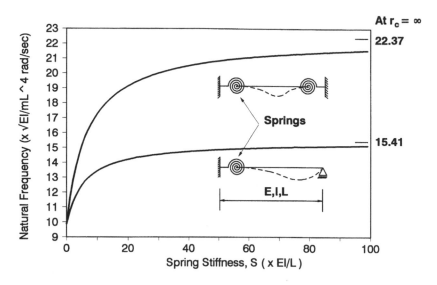

Figure 10. Natural frequency of beam with various spring stiffness.

is zero or infinite, the solutions converge to the analytical results obtained in standard texts [Paz (1980)]. When the joint stiffness is varied due to the unavoidable joint flexibility, the natural frequency changes and the result is depicted in Figure 10.

It can be seen in Figure 10 that the change of the natural frequency is considerable when the joint stiffness changes. Consequently, ignoring of this joint flexibility may generally lead to an unacceptable error in the analysis of these structures.

1.11.3. Vibration Analysis of a Cantilever Column Subjected to a Varying Axial Load

The presence of an axial load affects the stiffness of a beam-column. This effect can be included via the geometric stiffness matrix or by the stability function. It, in turn, affects the dynamical characteristics of a beam-column.

This example is aimed at the study of the effect of an axial load on the vibration of a cantilever column shown in Figure 11. In the figure, the computed frequency is expressed in a non-dimensional form via the division by the natural frequency, ω_o of the cantilever column free from any initial axial load. P_e is the Euler's buckling load of the cantilever. As can be seen in

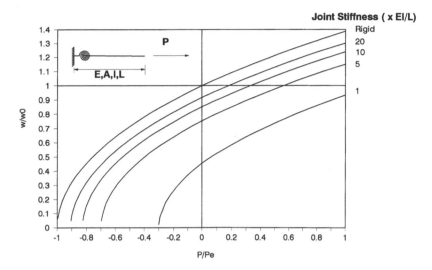

Figure 11. Natural frequency of a cantilever column with various axial load.

the figure, an axial force alters the natural frequency significantly when the force is large. If the axial load is in tension, the frequency of the member will be increased and vice versa. When the load is close to the Euler's buckling load, the frequency tends to zero, indicating an unstable equilibrium state.

1.11.4. Vibration of Two Storey Portals

The natural frequency of the two storey portal frame of height 7.32 m (288 inches) and width 6.10 m (240 inches) as shown in Figure 12 is analyzed. To activate the nonlinear characteristic of the joints, a set of point loads of P equal to 1601 kN (360 kips) is applied at the mid-span and the two ends of the beam. This arrangement will create moments at the two ends of the beam, which will, in turn change the joint stiffness and thus the frequency. Totally, three cases of rigid floor, unbraced and braced frames are studied in this example. In the rigid floor case, the slab is assumed to be hundred times stiffer than the column rigidity. Therefore, the flexural rigidity of the beams is strengthened by the slab. The results of the analysis are also tabulated in Table 2. The linear and nonlinear joint stiffness and geometry have been assumed and their results are presented. Totally, five types of joints are considered, namely they are the single web angle connection (A), the top and seated angle connection with double web cleats (B), the flush end plate connection (C), the extended end

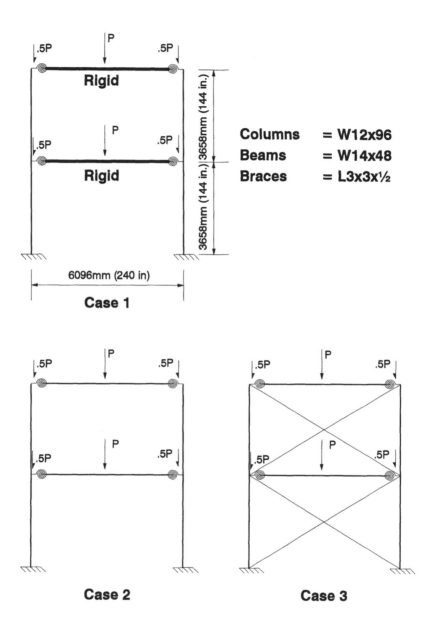

Figure 12. Vibration analysis of portal frames.

Table 2. Natural Frequency of One Bay Two Storey Frame with Semi-Rigid Connections

Joint Type	Natural Frequency, $f(Hz)$		Natural Frequency, $f(Hz)$ (with axial effect)	
	Linear Joint	Nonlinear Joint	Linear Joint	Nonlinear Joint
Case 1: Without Bracing (Rigid Floor Assumption)				
A	6.57	6.57	4.32	4.33
B	7.49	7.50	5.26	5.28
C	7.73	7.73	5.51	5.50
D	9.94	9.94	7.60	7.60
Rigid	17.87		14.58	
Case 2: Without Bracing				
A	6.44	5.26	4.19	2.80
B	7.13	5.27	4.91	2.82
C	7.30	5.29	5.08	2.84
D	8.53	5.29	6.28	2.85
Rigid	10.35		7.99	
Case 3: With Bracing				
A	26.38	26.31	21.63	21.57
B	26.44	26.31	21.68	21.57
C	26.45	26.31	21.69	21.57
D	26.58	26.31	21.79	21.57
Rigid	26.85		22.05	

plate connection (D) and the rigid joints. These typical joint rigidities are adopted from Lui and Chen (1988) and shown in Figure 4.

Generally speaking, joint flexibility affects the vibration characteristics of a structure through its influence on the structural stiffness. It can also be seen in Table 2 that the nonlinearity of joint stiffness may or may not have a notable influence over the vibration characteristics of a portal. The effects depend on the change of joint stiffness due to the applied loads. For example, if the initial stiffness before loading differs significantly from the stiffness after loading, the frequency will be altered and vice versa.

The effect of axial load on the natural frequency of a structure depends on the magnitude of the change of the member stiffness by the load and again cannot be over-generalized. For slender structures subjected to high axial load, this effect is significant and vice versa. In general, if columns of medium slenderness are designed without the consideration of axial load effect, the frequency of a structure will be significantly over-estimated.

Generally speaking, the influence of axial effects and joint flexibility can be inspected by studying the relationship between various matrices in Equation (61). However, if a more precise prediction of the vibration characteristics of a frame is required, a numerical analysis such as that presented in this chapter should be invoked.

1.11.5. The Effect of Semi-rigid Connections in the Seismic Design of Buildings

The problem of earthquake loading and response may be unique among the design criteria applied to buildings. The most common method of analysis used in seismic design is to replace the inertia loads acting on the structures by the so-called equivalent static seismic forces. This approach has been adopted in most of the design codes such as Uniform Building Codes and many others national codes. As the equivalent static forces depend on the natural frequency of the structure, an accurate method for calculating the structural frequency is needed.

Earthquake resistance requires energy absorption which means the structures should have predictable ductility as well as strength. The required ductility can be achieved by proper choice of framing and connection details. This example is to demonstrate the use of semi-rigid connections in seismic design of buildings.

The selected multi-storey frame is modified from the user manual of DYNAMIC/EASE2 and shown in Figure 13. The frame is detailed with rigid, extended end plate and flush end plate connections. Using the proposed method, the periods of the frame are determined and the results are listed in Table 3. It was found that the lateral drift of frame with flush end plate connection exceeds the allowable limit of H/500, although the equivalent static loads are the least in the considered cases.

Although the lateral deflection of the frame with extended end plate connections is larger than rigid frame, the elastic lateral deflection is still within the code allowable. In addition, the extra cost of providing fully rigid connections for a moment-resisting frame is eliminated.

1.12. CONCLUSIONS

The natural frequency of steel frames is dependent on the stiffness of the structures. In this chapter, two major nonlinearities that affect the stiffness of the steel frame structures have been discussed. They are nonlinear joint flexibility and geometric effects due to the presence of axial load.

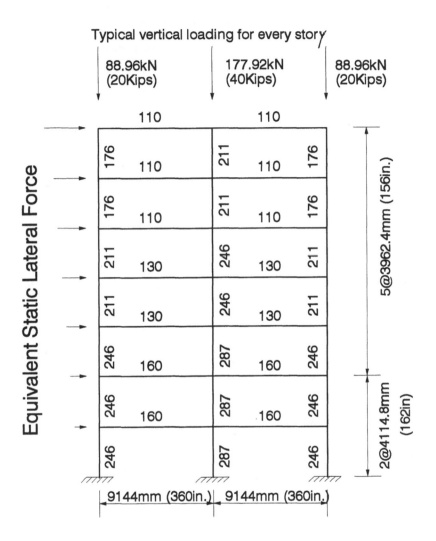

All columns are W14'S

All beams are W24'S

Member weights are indicated adjacent to the members

Figure 13. Two bays and seven storey frame.

Table 3. Seismic Design of Multi-Storey Frame with Semi-Rigid Connections

Coefficient	Type of Connections		
	Rigid	Extended End Plate	Flush End Plate
Period, T (sec)	1.26	2.66	3.7
Seismic Coefficient, C	5.93×10^{-2}	4.08×10^{-2}	3.47×10^{-2}
Site-Structure resonance factor, S	1.43	1.50	1.38
Base shear, V (kN/kips)	120.89/27.18	90.6/20.37	70.99/15.96
Top storey drift, Δ (mm/in.)	13.72/0.54	46.74/1.84	74.422/2.93

Note:
UBC formula : V=ZIKCSW
Z : Zone coefficient = 0.375 (Zone 2) in this example.
I : Importance factor= 1.0
K : Structural System Coefficient = 0.67
W : Weight of the structure = 5889kN (1324 kips)

In practical design, the natural frequency of a structure is obtained from the non-trivial solution of the mass and elastic stiffness matrices. However, with the present of axial load, the structures would either softened or hardened which would affect the stiffness and hence the structural frequency.

Apart from the geometrically nonlinear effect, joint flexibility is also a major source of nonlinearity that affect the natural period of steel frames. Because connection flexibility is described by the M-ϕ curve which is generally a nonlinear curve, special techniques in handling such behavior is required. A numerical procedure for free vibration analysis of flexibly jointed steel frames has been described. The proposed method adopts a shape function with springs at both ends. Therefore, the principle of total potential energy can be applied directly to obtain the element matrices. This technique is believed to be more general than the existing methods and does not involve complex mathematical manipulations.

As both geometric effect and joint flexibility may occur in the same structure, a procedure for nonlinear vibration analysis of structural steel frames is proposed. The presented method allows the incorporation of joint flexibility and consideration of initial stress in obtaining the natural periods of the steel structure. Examples of linear and nonlinear free vibration analyses of flexibly jointed frames are given to illustrate the application and accuracy of the suggested technique.

ACKNOWLEDGEMENT

The authors wish to express their thanks to financial support from the Hong Kong Polytechnic University. The encouragement by Prof. Michael Anson of Hong Kong Polytechnic University and Dr. Andrew K.C. Chan of the Ove Arup and Partners (HK) Ltd are greatly appreciated. Finally, the authors are thankful to Dr David Vesey for his invaluable comments on this manuscript.

References

1. Al-Bermani, F.G.A. and Kitipornchai, S. 1988, *Nonlinear analysis of thin-walled structures using one element per member*, Res. Report No. CE97, Dept. Civ. Engrg., Queensland University, Australia.

2. Ang, K.M. and Morris, G.A. 1984, "Analysis if Three-dimensional Frames with Flexible Beam-column Connections", *Canadian J. Civil Engineering*, **11**, 245–54.

3. Attiogbe, E. and Morris, G. 1991, "Moment-Rotation Functions for Steel Connections", *J. Structural Division*, ASCE, **117**(6), 1703–1718.

4. Azizinamini, A., Bradburn, J.H. and Radziminski, J.B. 1987, "Initial Stiffness of Semi-rigid Steel Beam-to Column Connections", *J. Constructional Steel Research*, **8**, 71–91.

5. Bathe, K.J., *Finite Element Procedures in Engineering Analysis*, Prentice-Hall Inc., Englewood Cliffs, New Jersey, U.S.A., 1982.

6. Bathe, K.J. and Bolourchi, S. 1979, "Large Displacement Analysis of Three-dimensional Beam Structures", *Int. J. Numerical Methods in Engineering*, **14**, 961–986.

7. Bathe., K.J. and Wilson, R.H. 1973, "NONSAP - A General Finite Element Program for Nonlinear Dynamic analysis of Complex Structures", Paper M3/1, *Proc. 2nd Int. Conference Structural Mechanics in Reactor Technology*, Berlin.

8. Batho, C and Rowan, H.C. 1934, *Investigation on beam and stanchion connections*, 2nd Report of Steel Structures, Her Majesty's Stationery Office, London, England.

9. Batoz, L.J. and Dhatt, G.(1979, "Incremental displacement algorithms for nonlinear problems", *Int. J. Num. Meth. in Engrg.*, **14**, 1262–1267.

10. Bell, W.G., Chesson, E. Jr. and Munse, W.H. 1958, *Static Tests of Standard Riveted and Bolted Beam-to-Column Connections*, University of Illinois Engineering Experiment Station, Urbana, Illinois.

11. Bennetts, I.D., Thomas, I.R. and Grundy, P. 1978, "Shear Connections for Beams to Columns", *Proceedings of the Conference on Metal Structures*, The Institution of Engineers, Perth, Australia, Preprints, pp. 102–106.

12. Bose, S.K., McNeice, G.M. and Sherbpurne, A.N. 1972, "Columns Webs in Steel Beam to Column Connections, Part I, Formulation and Verification", *Computers & Structures*, **2**, 253–272.

13. Carpenter, L.D. and Lu, L.W. 1973, *Reversed and Repeated Load Tests of Full-Scale Steel Frames*, Bulletin No. 24, AISC, New York, April.

14. Chan, S.L. 1988, "Geometric and material non-linear analysis of beam-columns and frames using the minimum residual displacement method", *Int. J. Num. Meth. in Engrg.*, **26**, 2657–2699.

15. Chan, S.L. 1989, "Inelastic post-buckling analysis of tubular beam-columns and frames", *Engrg. Struct.*, **110**, 23–30.

16. Chan, S.L. and Ho, W.M.G. 1994, "Nonlinear vibration analysis of steel frames with semi-rigid connections", *J. Struct. Div., ASCE*, **120**(4), 1075–1087.

17. Chan, S.L. and Kitipornchai, S. 1986, "Geometric Nonlinear Analysis of Asymmetric Thin-walled Beam-columns", *Engineering Structures*, **9**, 243–254.

18. Chmielowiec, M. and Richard, R.M. 1987, *Moment Rotation Curves for Partially Restrained Steel Connections*, Report to AISC, University of Arizona.

19. Clough, R.W. and Tang, D. 1975, *Earthquake Simulator Story of Steel Frame Structure, Vol. I : Experimental Results*, EERC Report 75–6, Earthquake Engineering Research Centre, University of California, Berkeley.

20. Colson, A. and Louveau, J.M. 1968, "Connections Incidence on the Elastic Behavior of Steel Structures", *Euromech Colloquium 174 on Inelastic Structures under Variable Loads*, Palermo, October, pp. 257–268.

21. Frye, M.J. and Morris, G.A. 1975, "Analysis of Flexibly Connected Steel Frames", *Canadian J. Civil Engineering*, **2**, 280–291.

22. Goto, Y. and Chen, W.F. 1987, "On the computer-based design analysis for the flexibly jointed frames", *J. Const. Steel Res.*, **8**, 203–231.

23. Goverdhan, A.V. 1934, *A Collection of Experimental Moment-Rotation Curves and Evaluation of Predicting Equations forSemi-Rigid Connections*, Master Thesis, Vanderbilt University, Nashville, Tennessee.

24. Grundy, P., Murray, N.W. and Bennetts, I.D., 1983, "Torsional Rigidity of Standard Beam-to-column Connections", *Proceedings of the Conference on Metal Structures*, The Institution of Engineers, Australia, pp. 164–169.

25. Ho, W.M.G. and Chan S.L. 1991, "Semi-bifurcation and bifurcation analysis of flexibly connected steel frames", *J. Struct. Div.*, ASCE, **117**(8), 2299–2319.

26. Ho, W.M.G. and Chan, S.L. 1993, " An accurate and efficient method for large deflection inelastic analysis of frames with semi-rigid connections", *J. Construct. Steel Research*, **26**, 171–191.

27. Huang, J.Y. 1958, *The Derivation of an Elastic Restraint Equation for the split Beam Semi-Rigid Beam-Column Connection*, Thesis, College of Engineering, Oklahoma State University.

28. Johnson, S.W., Law, C.L.C. 1981, "Semi-rigid Joints for composite frames", *Joints in Structural Steelwork* (ed. J.H. Howlett *et al*. Pentech Press, London, pp. 5.73–5.87.

29. Jones, S.W., Kirby, P.A. and Nethercot, D.A. 1982, "Columns with Semirigid Joints", *J. Structural Division*, ASCE, **108**(ST2), 361–372.

30. Jones, S.W., Kirby, P.A. and Nethercot, D.A. 1983, "The Analysis of Frames with Semi-Rigid Connections — A State-of-the-Art-Report", *J. of Constructional Steel Research*, **3**(2), 2–13.

31. Kawashima, S. and Fujimoto, T. 1984, "Vibration Analysis of Frames with Semi-rigid Connections", *Computer & Structures*, **19**, 85–92.

32. Kenndy, D.R.J. and Hafez, M. 1984, "A study of End-plate Connections for Steel Beams", *Canadian J. Civil Engineering*, **11**(2), 139–149.

33. Kishi, N. and Chen, W.F. 1986, *Data base of Steel beam-to-column Connections*, CE-STR-86-26, School of Civil Engineering, Purdue University, West Lafayette, Ind., USA.

34. Kishi, N. and Chen, W.F. 1987, *Moment-rotation Relations of Semi-rigid Connections*, Research Report CE-STR-87-29, School of Civil Engineering, Purdue University, Sept., p. 25.

35. Krawinker, H., Berterom V.V. and Popov, E.P. 1971, *Inelastic Behavior of Steel Beam-to-Column Sub-assemblages*, EERC Report 71-7, Earthquake Engineering Research Centre, University of California, Berkeley.

36. Krishnamurthy, N., Huang, H.T., Jeffrey, P.K. and Avery, L.K. 1979, "Analytical M-θ Curves for End-plate Connections", *J. Structural Division*, ASCE, **105**(ST1), 133–145.

37. Kukreti, A.R., Murray, T.M., Abolmaali, A. 1987, "End Plate Connection Moment-rotation Relationship", *J. Constructional Steel Research*, **8**, 137–157.

38. Lewitt, C.W., Chesson, E, Jr. and Munse, W.H. 1966, *Restraint Characteristics of Flexible Riveted and Bolted-to-column Connections*, Structural Research Series No. 296., Department of Civil Engineering, University of Illinois, Urbana, Illinois.

39. Lionberger, S.R. and Weaver, W. 1969, "Dynamic Response of Frames with Nonrigid Connections", *J. Engineering Mechanics Division, ACSE*, **95**(EM1), 95–114.

40. Lothers, J.E. 1951, "Elastic Restraints Equations for Semi Rigid Connections", *Transactions*, ASCE, **116**, 480–494.

41. Lothers, J.E. 1960, *Advances Design in Structural Steel*, Prentice Hall Inc., Englewood Ciffs, N.J., USA, pp. 390–391.

42. Lui, E.M. 1985, *Effects of Connection Flexibility and Panel Zone Deformation on the Behavior of Plane Steel Frames*, Ph.D. Dissertation, School of Civil Engineering, Purdue University, West Lafayette, Indiana.
43. Lui, E.M. and Chen, W.F. 1987a, "Effects of joint flexibility on the behavior of steel frames", *Comp. Struct.*, **26**(5), 719–732.
44. Lui, E.M. and Chen, W.F. 1987b, "Steel frame analysis with flexible joints", *J. Construct. Steel Research*, **8**, 161–202.
45. Lui, E.M. and Chen W.F. 1988, "Behavior of Braced and Unbraced Semi-rigid Frames", *Int. J. Solids Structures*, **24**(9), 893–913.
46. Morris, G.A. and Packer,J.A. 1987, "Beam-to-column Connections in Steel Frames", *Canadian J. Civil Engineering*, **14**, 68–76.
47. Munse, W.H., Bell, W.G. and Chesson, E. 1959, "Behavior of Riveted and Bolted Beam-to-Column Connections", *J. Structural Division*, ASCE, **85**(ST3), 25–50.
48. Nee, K.M. and Haldar, A. 1988, "Elastoplastic nonlinear post-buckling analysis of partially restrained space structures", *Comp. Meth. in A Mech. Engrg.*, **71**, 69–97.
49. Nethercot, D.A. 1985, *Steel Beam to Column Connections — A Review of Test Data and their Application to the Evaluation of Joint Behavior on the Performance of Steel Frames*, CIRIA Project Record 338, London.
50. Nethercot, D.A., Kirby, P.A. and Rifal, A.M. 1987, "Columns in Partially Restrained Construction : Analytical Studies", *Canadian J. Civil Engineering*, **14**(4), 485–497.
51. Oden, J.T. and Reddy, J.N. 1976, *Variational Methods in Theoretical Mechanics*, Springer-Verlag, Berlin, Germany.
52. Patel, K.V. and Chen, W.F. 1984, "Nonlinear Analysis of Steel Moment Connections", *J. Structural Engineering*, ASCE, **110**(8), 1861–1874.
53. Paz, M. 1980, *Structural Dynamics : Theory and Computation*, Van Nostrand Reinhold Ltd., Canada.
54. Pogg, C. 1988, "A finite element model for the analysis of flexibility connected steel frames", *Int. J. Numer. Meth. Engrg.*, **26**, 2239–2254.
55. Oran, C. 1973, "Tangent stiffness in space frames", *J. Struct. Div., ASCE*, **99**(ST6), 987–1001.
56. Popov, E.P. and Pinkey, R.B. 1967a, *Behavior of Steel Building Connections Subjected to Inelastic Strain Reversal*, SESM Report 67–30, Structures and Material Research, Department of Civil Engineering, University of California, Berkeley.
57. Popov, E.P. and Pinkey, R.B. 1967b, *Behavior of Steel Building Connections Subjected to Inelastic Strain Reversal–Experimental Data*, SESM Report 67–31, Structures and Material Research, Department of Civil Engineering, University of California, Berkeley.
58. Popov, E.P. and Pinkey, R.B. 1969, "Cyclic Yield Reversal in Steel Building Joint", *J. Structural Division*, ASCE, **95**(ST3), 327–353.
59. Popov, E.P. and Stephen, R.M. 1970, *Cyclic Loading of Full-Size Steel Connections*, EERC Report 70–3, Earthquake Engineering Research Centre, University of California, Berkeley.
60. Popov, E.P. and Stephen, R.M. 1976, *Tensile Capacity of Partual Penetration Welds*, EERC Report 76–28, Earthquake Engineering Research Centre, University of California, Berkeley.
61. Richard, R.M. and Abbott, B.J. 1975, "Versatile Elastic-plastic Stress-strain Formula", *J. Engineering Mechanic Division*, ASCE, **101**(EM4), 511–515.
62. Ramberg, R.M. and Osgood, W.R. 1943, *Description of Stress-strain Curves by Three Parameters*, Technical Report No. 902, National Advisory Committee for Aeronautics, Washington, D.C.
63. Rathbun, J.C. 1936, "Elastic Properties of Riveted Connections", *Transactions*, ASCE, **101**, 524–563.
64. Reedy, J.N. 1984, *An Introduction to the Finite Element Method*, McGraw-Hill Book Company, New York, USA.
65. Reedy, J.N. 1986, *Applied Functional Analysis and Variational Methods in Engineering*, McGraw-Hill Book Company, New York, USA.
66. Shi, G. and Atluri, S.N. 1989, "Static and dynamic analysis of space frames with non-linear flexible connections", *Int. J. Num. Meths. in Engrg.*, **28**, 2635–2650.

67. Sivakumaran, K.S. 1988, "Seismic Response of Multi-storey Steel Buildings with Flexible Connections", *Engineering Structures*, **10**, 238–288.
68. So, A.K.W. and Chan, S.L.(1991, "Buckling Analysis of Frames Using 1 Element/Member", *J. Construct. Steel Research*, **20**, 271–289.
69. Sommer, W.H. 1969, *Behavior of Welded Header Plate Connections*, Master Thesis, University of Toronto, Ontario, Canada.
70. Washizu, K., 1982, *Variational Methods in Elasticity and Plasticity*, 3rd Edition, Pergamom Press, Oxford, UK.
71. Wales, M.W. and Rossow, R.E. 1983, "Coupled Moment-axial Force Behavior in Bolted Joints", *J. Structural Division*, ASCE, **109**(5), 1250–1266.
72. Wilson, W.M. and Moore, H.F. 1917, *Tests to Determine the Rigidity of Riveted Joints in Steel Structures*, University of Illinois, Engineering Experiment Station, Bulletin No. 104, Urbana, USA.
73. Wu, F.H. and Chen, W.F. 1990, "A Design Model for Semi-rigid Connections", *Engineering Structures*, **12**(4), 88–97.
74. Yang Y.B. and McGuire, W. 1985, "A work control method for geometrically nonlinear analysis", *Proc. of the NUMETA '85 Conf.*, Elsevier Science Publishers Ltd., England, 913–921.
75. Yee, Y.L. and Melchers, R.E. 1986, "Moment-rotation Curves for Bolted Connections", *J. Structural Engineering*, ASCE, **112**(3), 615–635.
76. Young, C.R. and Jackson, K.B. 1934, "The Relative Rigidity of Weld and Riveted Connections", *Canadian J. Research*, **2**(1, 62–100 and **2**(2), 101–134.
77. Yu, C.H. and Shanmugam, N.E. 1986, "Stability of frames with semi-rigid joints", *Comp. Struct.*, **23**(5), 639–648.
78. Yu, S.Y. 1953, *The Derivation of an Equation for the Elastic Restraint of the Top and Seated Angle Type of Semi-Rigid Connection*, Thesis, College of Engineering, Oklahoma State University.
79. Yu, W.W. 1955, *The Derivation of an Elastic Restraint Equation for Combined Top and Seat with Web Angle Semi-Rigid Beam-Column Connection*, Thesis, College of Engineering, Oklahoma State University.
80. Zienkiewicz, O.C. and Taylor, R.L. 1989, *The Finite Element Method*, 4th Edition, McGraw-Hill International Editions, London, UK.
81. American Institute of Steel Construction 1986, *Load and Resistance Factor Design Specification for structural steel buildings*, September, Chicago, USA, AISC.
82. British Standard Institution 1985, *BS5950: Part 1: 1985, Structural use of steelwork in building*, London, BSI.
83. *Uniform Building Codes*, 1985, ICBO.
84. *Static and Dynamic Analysis of Multistory Frame Using DYNAMIC/EASE2*, Engineering Analysis Corporation and Computers/Structures International.

1.13. APPENDICES

APPENDIX 1: The 12 × 12 Linear Stiffness Matrix, [KL]

The 12×12 Linear Stiffness Matrix, $[KL]$ is symmetric. The non-zero terms in the upper triangle (in local co-ordinates) are given below.

$$KL_{1,1} = EA/L$$

$$KL_{1,7} = -EA/L$$

$$KL_{2,2} = KL_{11}^z = EI_z\{3(C_{kl}^z)^{-1}((384(S_i^z)^2 + 192k_{11}^z S_i^z + 96(k_{11}^z)^2)(S_j^z)^2$$
$$192k_{11}^z(S_i^z)^2 - 96(k_{11}^z)^2 S_i^z)S_j^z + 96(k_{11}^z)^2(S_i^z)^2)$$

$$KL_{2,6} = KL_{12}^y = EI_z\{2(C_{kl}^z)^{-1}(192(S_i^z)^2(S_j^z)^2 + 192k_{11}^z(S_i^z)^2$$
$$- 48(k_{11}^z)^2 S_i^z)S_j^z + 96(k_{11}^z)^2(S_i^z)^2)$$

$$KL_{2,8} = KL_{13}^y = EI_z\{3(C_{kl}^z)^{-1} - ((384(S_i^z)^2 + 192k_{11}^z S_i^z + 96(k_{11}^z)^2)(S_j^z)^2$$
$$+ 192k_{11}^z(S_i^z)^2 - 96(k_{11}^z)^2 S_i^z)S_j^z + 96(k_{11}^z)^2(S_i^z)^2)$$

$$KL_{2,12} = KL_4^y = EI_z\{2(C_{kl}^z)^{-1}((192(S_i^z)^2 + 192k_{11}^z S_i^z + 96(k_{11}^z)^2)(S_j^z)^2$$
$$- 48(k_{11}^z)^2 S_i^z S_j^z)$$

$$KL_{3,3} = KL_{11}^y = EI_y\{3(C_{kl}^y)^{-1}((384(S_i^y)^2 + 192k_{11}^y S_i^y + 96(k_{11}^y)^2)(S_j^y)^2$$
$$+ 192k_{11}^y(S_i^y)^2 - 96(k_{11}^y)^2 S_i^y)S_j^y + 96(k_{11}^y)^2(S_i^y)^2)$$

$$KL_{3,5} = KL_{12}^y = -EI_y\{2(C_{kl}^y)^{-1}(192(S_i^y)^2(S_j^y)^2$$
$$+ 192k_{11}^y(S_i^y)^2 - 48(k_{11}^y)^2 S_i^y)S_j^y + 96(k_{11}^y)^2(S_i^y)^2)$$

$$KL_{3,9} = KL_{13}^y = EI_y\{3(C_{kl}^y)^{-1} - ((384(S_i^y)^2 + 192k_{11}^y S_i^y + 96(k_{11}^y)^2)(S_j^y)^2$$
$$+ 192k_{11}^y(S_i^y)^2 - 96(k_{11}^y)^2 S_i^y)S_j^y + 96(k_{11}^y)^2(S_i^y)^2)$$

$$KL_{3,11} = KL_{14}^y = -EI_y\{2(C_{kl}^y)^{-1}((192(S_i^y)^2 + 192k_{11}^y S_i^y + 96(k_{11}^y)^2)(S_j^y)^2$$
$$- 48(k_{11}^y)^2 S_i^y S_j^y)$$

$$KL_{4,4} = GJ/L$$

$$KL_{4,7} = -GJ/L$$

$$KL_{5,5} = KL_{22}^y = EI_y\{1(C_{kl}^y)^{-1}(128(S_i^y)^2(S_j^y)^2 + 192k_{11}^y(S_i^y)^2 S_j^y +$$
$$96(k_{11}^y)^2(S_i^y)^2)$$

$$KL_{5,9} = KL_{23}^y = -EI_y\{2(C_{kl}^y)^{-1} - (192(S_i^y)^2(S_j^y)^2 + 192k_{11}^y(S_i^y)^2$$
$$- 48(k_{11}^y)^2 S_i^y)S_j^y + 96(k_{11}^y)^2(S_i^y)^2)$$

$$KL_{5,11} = KL_{24}^y = EI_y\{1(C_{kl}^y)^{-1}(64(S_i^y)^2(S_j^y)^2 - 48(k_{11}^y)^2 S_i^y S_j^y)$$

$$KL_{6,6} = KL_{22}^z = EI_z\{1(C_{kl}^z)^{-1}(128(S_i^z)^2(S_j^z)^2 + 192k_{11}^z(S_i^z)^2 S_j^z$$
$$+ 96(k_{11}^z)^2(S_i^z)^2)$$

$$KL_{6,8} = KL_{23}^z = EI_z\{_2(C_{kl}^z)^{-1} - (192(S_i^z)^2(S_j^z)^2 + 192k_{11}^2(S_i^z)^2$$
$$- 48(k_{11}^z)^2 S_i^z)S_j^z + 96(k_{11}^z)^2(S_j^z)^2)$$

$$KL_{6,12} = KL_{24}^z = EI_z\{_1(C_{kl}^z)^{-1}(64(S_i^z)^2(S_j^z)^2 - 48(k_{11}^z)^2 S_i^z S_j^z)$$

$$KL_{7,7} = EA/L$$

$$KL_{8,8} = KL_{33}^z = EI_z\{_3(C_{kl}^z)^{-1}((384(S_i^z)^2 + 192k_{11}^z S_i^z + 96(k_{11}^z)^2)(S_j^z)^2$$
$$+ 192k_{11}^z(S_i^z)^2 - 96(k_{11}^z)^2 S_i^z)S_j^z + 96(k_{11}^z)^2(S_i^z)^2)$$

$$KL_{8,12} = KL_{34}^z = EI_z\{_2(C_{kl}^z)^{-1} - ((192(S_i^z)^2 + 192k_{11}^z S_i^z + 96(k_{11}^z)^2)(S_j^z)^2$$
$$- 48(k_{11}^z)^2 S_i^z S_j^z)$$

$$KL_{9,9} = KL_{33}^y = EI_y\{_3(C_{kl}^y)^{-1}((384(S_i^y)^2 + 192k_{11}^y S_i^y + 96(k_{11}^y)^2)(S_j^y)^2$$
$$+ 192k_{11}^y(S_i^y)^2 - 96(k_{11}^y)^2 S_i^y)S_j^y + 96(k_{11}^y)^2(S_i^y)^2)$$

$$KL_{9,11} = KL_{34}^y = -EI_y\{_2(C_{kl}^y)^{-1} - ((192(S_i^y)^2 + 192k_{11}^y S_i^y$$
$$+ 96(k_{11}^y)^2)(S_j^y)^2 - 48(k_{11}^y)^2 S_i^y S_j^y)$$

$$KL_{10,10} = GJ/L$$

$$KL_{11,11} = KL_{44}^y = EI_y\{_1(C_{kl}^y)^{-1}(128(S_i^y)^2 + 192k_{11}^y S_i^y + 96(k_{11}^y)^2)(S_j^y)^2$$

$$KL_{12,12} = KL_{44}^z = EI_z\{_1(C_{kl}^z)^{-1}(128(S_i^z)^2 + 192k_{11}^z S_i^z + 96(k_{11}^z)^2)(S_j^z)^2$$

where

$$k_{11} = 4EI_z/L$$

$$k_{11} = 4EI_y/L$$

$$_1(C_{kl}^y)^{-1} = 2((16L(S_i^y)^2 + 32k_{11}^y LS_i^y + 16(k_{11}^y)^2 L)(S_j^y)^2 + (32k_{11}^y L(S_i^y)^2$$
$$+ 56(k_{11}^y)^2 LS_i^y + 24(k_{11}^y)^3 L)S_j^y + 16(k_{11}^y)^2 L(S_i^y)^2$$
$$+ 24(k_{11}^y)^3 LS_i^y + 9(k_{11}^y)^4 L)$$

$$_2(C_{kl}^y)^{-1} = 2((16L^2(S_i^y)^2 + 32k_{11}^y L^2 S_i^y + 16(k_{11}^y)^2 L^2)(S_j^y)^2$$
$$+ (32k_{11}^y L^2(S_i^y)^2 + 56(k_{11}^y)^2 L^2 S_i^y + 24(k_{11}^y)^3 L^2)S_j^y$$
$$+ 16(k_{11}^y)^2 L^2(S_i^y)^2 + 24(k_{11}^y)3L2S_i^y + 9(k_{11}^y)^4 L^2)$$

$$_3(C_{kl}^y)^{-1} = 2((16L^3(S_i^y)^2 + 32k_{11}^y L^3 S_i^y + 16(k_{11}^y)^2 L^3)(S_j^y)^2$$
$$+ (32k_{11}^y L^3(S_i^y)^2 + 56(k_{11}^y)^2 L^3 S_i^y + 24(k_{11}^y)^3 L^3)S_j^y$$
$$+ 16(k_{11}^y)^2 L^3(S_i^y)^2 + 24(k_{11}^y)^3 L^3 S_i^y + 9(k_{11}^y)^4 L^3)$$

$$_1(C_{kl}^z)^{-1} = 2((16L(S_i^z)^2 + 32k_{11}^z LS_i^z + 16(k_{11}^z)^2 L)(S_j^z)^2$$
$$+ (32k_{11}^z L(S_i^z)^2 + 56(k_{11}^z)^2 LS_i^z + 24(k_{11}^z)^3 L)S_j^z$$
$$+ 16(k_{11}^z)^2 L(S_i^z)^2 + 24(k_{11}^z)^3 LS_i^z + 9(k_{11}^z)^4 L)$$

$$_2(C_{kl}^z)^{-1} = 2((16L^2(S_i^z)^2 + 32k_{11}^z L^2 S_i^z + 16(k_{11}^z)^2 L^2)(S_j^z)^2$$
$$+ (32k_{11}^z L^2(S_i^z)^2 + 56(k_{11}^z)^2 L^2 S_i^z + 24(k_{11}^z)^3 L^2)S_j^z$$
$$+ 16(k_{11}^z)^2 L^2(S_i^z)^2 + 24(k_{11}^z)3L2S_i^z + 9(k_{11}^z)^4 L^2)$$

$$_3(C_{kl}^z)^{-1} = 2((16L^3(S_i^z)^2 + 32k_{11}^z L^3 S_i^z + 16(k_{11}^z)^2 L3)(S_j^z)^2$$
$$+ (32k_{11}^z L^3(S_i^z)^2 + 56(k_{11}^z)^2 L^3 S_i^z + 24(k_{11}^z)^3 L^3)S_j^z$$
$$+ 16(k_{11}^z)^2 L^3(S_i^z)^2 + 24(k_{11}^z)^3 L^3 S_i^z + 9(k_{11}^z)^4 L^3)$$

APPENDIX 2: The 12×12 Geometric Stiffness Matrix, [KG]

The 12×12 Geometric Stiffness Matrix, $[KG]$ is symmetric. The non-zero terms in the upper triangle (in local co-ordinates) are given below.

$$KG_{1,1} = 0$$

$$KG_{2,2} = KG_{11}^z = P\{_1(C_{kl}^z)^{-1}((192(S_i^z)^2 + 336k_{11}^z S_i^z + 192(k_{11}^z)^2)(S_j^z)^2$$
$$+ (336k_{11}^z(S_i^z)^2 + 504(k_{11}^z)^2 S_i^z + 240(k_{11}^z)^3)S_j^z$$
$$+ 192(k_{11}^z)^2(S_i^z)^2 + 240(k_{11}^z)^3 S_i^z + 90(k_{11}^z)^4)\}$$

$$KG_{2,6} = KG_{12}^z = P\{_2(C_{kl}^z)^{-1}((16(S_i^z)^2 - 16k_{11}^z S_i^z)(S_j^z)^2$$
$$+ (32k_{11}^z(S_i^z)^2 - 28(k_{11}^z)^2 S_i^z)S_j^z + 32(k_{11}^z)^2(S_i^z)^2)\}$$

$$KG_{2,8} = KG_{13}^z = P\{_1(C_{kl}^z)^{-1} - ((192(S_i^z)^2 + 336k_{11}^z S_i^z + 192(k_{11}^z)^2)(S_j^z)^2$$
$$+ (336k_{11}^z(S_i^z)^2 + 504(k_{11}^z)^2 S_i^z + 240(k_{11}^z)^3)S_j^z$$
$$+ 192(k_{11}^z)^2(S_i^z)^2 + 240(k_{11}^z)^3 S_i^z + 90(k_{11}^z)^4)\}$$

$$KG_{2,12} = KG_{14}^z = P\{_2(C_{kl}^z)^{-1}((16(S_i^z)^2 + 32k_{11}^z S_i^z + 32(k_{11}^z)^2)(S_j^z)^2$$
$$+ (-16k_{11}^z(S_i^z)^2 - 28(k_{11}^z)^2 S_i^z)S_j^z)\}$$

$$KG_{3,3} = KG_{11}^y = P\{_1(C_{kl}^y)^{-1}((192(S_i^y)^2 + 336k_{11}^y S_i^y + 192(k_{11}^y)^2)(S_j^y)^2$$
$$+ (336k_{11}^y(S_i^y)^2 + 504(k_{11}^y)^2 S_i^y + 240(k_{11}^y)^3)S_j^y$$
$$+ 192(k_{11}^y)^2(S_i^y)^2 + 240(k_{11}^y)^3 S_i^y + 90(k_{11}^y)^4)\}$$

$$KG_{3,5} = KG_{12}^y = -P\{2(C_{kl}^y)^{-1}((16(S_i^y)^2 - 16k_{11}^y S_i^y)(S_j^y)^2$$
$$+ (32k_{11}^y (S_i^y)^2 - 28(k_{11}^y)^2 S_i^y)S_j^y + 32(k_{11}^y)^2(S_i^y)^2)\}$$

$$KG_{3,9} = KG_{13}^y = P\{1(C_{kl}^y)^{-1} - ((192(S_i^y)^2 + 336k_{11}^y S_i^y + 192(k_{11}^y)^2)(S_j^y)^2$$
$$+ (336k_{11}^y (S_i^y)^2 + 504(k_{11}^y)^2 S_i^y + 240(k_{11}^y)^3)S_j^y$$
$$+ 192(k_{11}^y)^2(S_i^y)^2 + 240(k_{11}^y)^3 S_i^y + 90(k_{11}^y)^4)\}$$

$$KG_{3,11} = KG_{14}^y = -P\{2(C_{kl}^y)^{-1}((16(S_i^y)^2 + 32k_{11}^y S_i^y + 32(k_{11}^y)^2)(S_j^y)^2$$
$$+ (-16k_{11}^y (S_i^y)^2 - 28(k_{11}^y)^2 S_i^y)S_j^y)\}$$

$$KG_{4,4} = Pr^2/L$$

$$KG_{4,8} = P/L$$

$$KG_{4,9} = P/L$$

$$KG_{5,5} = KG_{12}^y = P\{3(C_{kl}^y)^{-1}(64L(S_i^y)^2(S_j^y)^2 + 144k_{11}^y L(S_i^y)^2 S_j^y$$
$$+ 96(k_{11}^y)^2 L(S_i^y)^2)$$

$$KG_{5,9} = KG_{12}^y = -P\{2(C_{kl}^y)^{-1} - ((16(S_i^y)^2 - 16k_{11}^y S_i^y)(S_j^y)^2$$
$$+ (32k_{11}^y (S_i^y)^2 - 28(k_{11}^y)^2 S_i^y)S_j^y + 32(k_{11}^y)^2(S_i^y)^2)\}$$

$$KG_{5,11} = KG_{14}^y = P\{3(C_{kl}^y)^{-1} - ((16L(S_i^y)^2 + 48k_{11}^y L S_i^y)(S_j^y)^2$$
$$+ (48k_{11}^y L(S_i^y)^2 + 84(k_{11}^y)^2 L S_i^y)S_j^y)\}$$

$$KG_{6,6} = KG_{12}^z = P\{3(C_{kl}^z)^{-1}(64L(S_i^z)^2(S_j^z)^2 + 144k_{11}^z L(S_i^z)^2 S_j^z$$
$$+ 96(k_{11}^z)^2 L(S_i^z)^2)\}$$

$$KG_{6,8} = KG_{12}^z = P\{2(C_{kl}^z)^{-1} - ((16(S_i^z)^2 - 16k_{11}^z S_i^z)(S_j^2)^2$$
$$+ (32k_{11}^z (S_i^z)^2 - 28(k_{11}^z)^2 S_i^z)S_j^z + 32(k_{11}^z)^2(S_i^z)^2)\}$$

$$KG_{6,12} = KG_{14}^z = P\{3(C_{kl}^z)^{-1} - ((16L(S_i^z)^2 + 48k_{11}^z L S_i^z)(S_j^z)^2$$
$$+ (48k_{11}^z L(S_i^z)^2 + 84(k_{11}^z)^2 L S_i^z)S_j^z)\}$$

$$KG_{7,7} = 0$$

$$KG_{8,8} = KG_{13}^z = P\{1(C_{kl}^z)^{-1}((192(S_i^z)^2 + 336k_{11}^z S_i^z + 192(k_{11}^z)^2)(S_j^z)^2$$
$$+ (336k_{11}^z (S_i^z)^2 + 504(k_{11}^z)^2 S_i^z + 240(k_{11}^z)^3)S_j^2$$
$$+ 192(k_{11}^z)^2(S_i^z)^2 + 240(k_{11}^z)^3 S_i^z + 90(k_{11}^z)^4)\}$$

$$KG_{8,12} = KG^z_{14} = P\{_1(C^y_{kl})^{-1} - ((16(S^z_i)^2 + 32k^z_{11}S^z_i + 32(k^z_{11})^2)(S^z_j)^2$$
$$+ (-16k^z_{11}(S^z_i)^2 - 28(k^z_{11})^2 S^z_i)S^z_j)\}$$

$$KG_{9,9} = KG^y_{13} = P\{_1(C^y_{kl})^{-1}((192(S^y_i)^2 + 336k^y_{11}S^y_i + 192(k^y_{11})^2)(S^y_j)^2$$
$$+ (336k^y_{11}(S^y_i)^2 + 504(k^y_{11})^2 S^y_i + 240(k^y_{11})^3)S^y_j$$
$$+ 192(k^y_{11})^2(S^y_i)^2 + 240(k^y_{11})^3 S^y_i + 90(k^y_{11})^4)\}$$

$$KG_{9,11} = KG^y_{14} = P\{_1(C^y_{kl})^{-1} - ((16(S^y_i)^2 + 32k^y_{11}S^y_i + 32(k^y_{11})^2)(S^y_j)^2$$
$$+ (-16k^y_{11}(S^y_i)^2 - 28(k^y_{11})^2 S^y_i)S^y_j)\}$$

$$KG_{10,10} = Pr^2/L$$

$$KG_{11,11} = KG^y_{14} = P\{_3(C^y_{kl})^{-1}(64L(S^y_i)^2 + 144k^y_{11}LS^y_i + 96(k^y_{11})^2 L)(S^y_j)^2\}$$

$$KG_{12,12} = KG^z_{14} = P\{_3(C^z_{kl})^{-1}(64L(S^z_i)^2 + 144k^z_{11}LS^z_i + 96(k^z_{11})^2 L)(S^z_j)^2\}$$

where

$$k^z_{11} = 4EI_z/L$$
$$k^y_{11} = 4EI_y/L$$

$$(_1C^z_{kg})^{-1} = 2((80L(S^z_i)^2 + 160k^z_{11}LS^z_i + 80(k^z_{11})^2 L)(S^z_j)^2$$
$$+ (160k^z_{11}L(S^z_i)^2 + 280(k^z_{11})^2 LS^z_i + 120(k^z_{11})^3 L)S^z_j$$
$$+ 80(k^z_{11})2L(S^z_i)^2 + 120(k^z_{11})^3 LS^z_i + 45(k^z_{11})^4 L)$$

$$(_2C^z_{kg})^{-1} = 2((80(S^z_i)^2 + 160k^z_{11}S^z_i + 80(k^z_{11})^2)(S^z_j)^2$$
$$+ (160k^z_{11}(S^z_i)^2 + 280(k^z_{11})^2 S^z_i + 120(k^z_{11})^3)S^z_j$$
$$+ 80(k^z_{11})^2(S^z_i)^2 + 120(k^z_{11})^3 S^z_i + 45(k^z_{11})^4)$$

$$(_3C^z_{kg})^{-1} = 2((240(S^z_i)^2 + 480k^z_{11}S^z_i + 240(k^z_{11})^2)(S^z_j)^2$$
$$+ (480k^z_{11}(S^z_i)^2 + 840(k^z_{11})^2 S^z_i + 360(k^z_{11})^3)S^z_j$$
$$+ 240(k^z_{11})^2(S^z_i)^2 + 360(k^z_{11})^3 S^z_i + 135(k^z_{11})^4)$$

$$(_1C^y_{kg})^{-1} = 2((80L(S^y_i)^2 + 160k^y_{11}LS^y_i + 80(k^y_{11})^2 L)(S^y_j)^2$$
$$+ (160k^y_{11}L(S^y_i)^2 + 280(k^y_{11})^2 LS^y_i + 120(k^y_{11})^3 L)S^y_j$$
$$+ 80(k^y_{11})2L(S^y_i)^2 + 120(k^y_{11})^3 LS^y_i + 45(k^y_{11})^4 L)$$

$$(_2C^y_{kg})^{-1} = 2((80(S^y_i)^2 + 160k^y_{11}S^y_i + 80(k^y_{11})^2)(S^y_j)^2$$
$$+ (160k^y_{11}(S^y_i)^2 + 280(k^y_{11})^2 S^y_i + 120(k^y_{11})^3)S^y_j$$
$$+ 80(k^y_{11})^2(S^y_i)^2 + 120(k^y_{11})^3 S^y_i + 45(k^y_{11})^4)$$

$$(_3C_{kg}^y)^{-1} = 2((240(S_i^y)^2 + 480k_{11}^y S_i^y + 240(k_{11}^y)^2)(S_j^y)^2$$
$$+ (480k_{11}^y(S_i^y)^2 + 840(k_{11}^y)^2 S_i^y + 360(k_{11}^y)^3)S_j^y$$
$$+ 240(k_{11}^y)^2(S_i^y)^2 + 360(k_{11}^y)^3 S_i^y + 135(k_{11}^y)^4)$$

APPENDIX 3: The 12 × 12 Consistent Mass Matrix, [M]

The 12 × 12 Consistent Mass Matrix, $[M]$ is symmetric. The non-zero terms in the upper triangle (in local co-ordinates) are given below.

$$M_{1,1} = 140\bar{m}L/420$$

$$M_{1,7} = 70\bar{m}L/420$$

$$M_{2,2} = M_{11}^z = \bar{m}\{(_1C_m^z)^{-1}[(416L(S_i^z)^2 + 656k_{11}^z LS_i^z + 264(k_{11}^z)^2L)(S_j^z)^2$$
$$+ (936k_{11}^z L(S_i^z)^2 + 1312(k_{11}^z)^2 LS_i^z + 462(k_{11}^z)^3L)S_j^z$$
$$+ 544(k_{11}^z)^2 L(S_i^z)^2 + 672(k_{11}^z)^3 LS_i^z + 210(k_{11}^z)^4L)]\}$$

$$M_{2,6} = M_{12}^z = \bar{m}\{(_2C_m^z)^{-1}[(176L^2(S_i^z)^2 + 128k_{11}^z L^2 S_i^z)(S)^2$$
$$+ (440k_{11}^z L^2(S_i^z)^2 + 288(k_{11}^z)^2 L^2 S_i^z)S_j^z + 288(k_{11}^z)^2 L^2(S_i^z)^2$$
$$+ 168(k_{11}^z)^3 L^2 S_i^z]\}$$

$$M_{2,8} = M_{13}^z = \bar{m}\{(_1C_m^z)^{-1}[(144L(S_i^z)^2 + 324k_{11}^z LS_i^z + 156(k_{11}^z)^2L)(S_j^z)^2$$
$$+ (324k_{11}^z L(S_i^z)^2 + 648(k_{11}^z)^2 LS_i^z + 273(k_{11}^z)^3L)S_j^z$$
$$+ 156(k_{11}^z)^2 L(S_i^z)^2 + 273(k_{11}^z)^3 LS_i^z + 105(k_{11}^z)^4L]\}$$

$$M_{2,12} = M_{14}^z = \bar{m}\{(_2C_m^z)^{-1}[-(104L^2(S_i^z)^2 + 260k_{11}^z L^2 S_i^z$$
$$+ 132(k_{11}^z)^2 L^2)(S_j^z)^2 + (152k_{11}^z L^2(S_i^z)^2 + 342(k_{11}^z)^2 L^2 S_i^z$$
$$+ 147(k_{11}^z)^3 L^2)S_j^z]\}$$

$$M_{3,3} = M_{11}^y = \bar{m}\{(_1C_m^y)^{-1}[(416L(S_i^y)^2 + 656k_{11}^y LS_i^y + 264(k_{11}^y)^2L)(S_j^y)^2$$
$$+ (936k_{11}^y L(S_i^y)^2 + 1312(k_{11}^y)^2 LS_i^y + 462(k_{11}^y)^3L)S_j^y$$
$$+ 544(k_{11}^y)^2 L(S_i^y)^2 + 672(k_{11}^y)^3 LS_i^y + 210(k_{11}^y)^4L)]\}$$

$$M_{3,5} = M_{12}^y = -\bar{m}\{(_2C_m^y)^{-1}[(176L^2(S_i^y)^2 + 128k_{11}^y L^2 S_i^y)(S)^2$$
$$+ (440k_{11}^y L^2(S_i^y)^2 + 288(k_{11}^y)^2 L^2 S_i^y)S_j^y + 288(k_{11}^y)^2 L^2(S_i^y)^2$$
$$+ 168(k_{11}^y)^3 L^2 S_i^y]\}$$

$$M_{3,9} = M_{13}^y = \bar{m}\{(_1C_m^y)^{-1}[(144L(S_i^y)^2 + 324k_{11}^y LS_i^y + 156(k_{11}^y)^2 L)(S_j^y)^2$$
$$+ (324k_{11}^y L(S_i^y)^2 + 648(k_{11}^y)^2 LS_i^y + 273(k_{11}^y)^3 L)S_j^y$$
$$+ 156(k_{11}^y)^2 L(S_i^y)^2 + 273(k_{11}^y)^3 LS_i^y + 105(k_{11}^y)^4 L]\}$$

$$M_{3,11} = M_{14}^y = -\bar{m}\{(_2C_m^y)^{-1}[-(104L^2(S_i^y)^2 + 260k_{11}^y L^2 S_i^y$$
$$+ 132(k_{11}^y)^2 L^2)(S_j^y)^2 + (152k_{11}^y L^2(S_i^y)^2 + 342(k_{11}^y)^2 L^2 S_i^y$$
$$+ 147(k_{11}^y)^3 L^2)S_j^y]\}$$

$$M_{4,4} = 140\bar{m}LJ/(420A)$$

$$M_{4,10} = 70\bar{m}LJ/(420A)$$

$$M_{5,5} = M_{22}^y = \bar{m}\{(_2C_m^y)^{-1}[32L^3(S_i^y)^2(S_j^y)^2 + 88k_{11}^y L^3(S_i^y)^2 S_j^y$$
$$+ 64(k_{11}^y)^2 L^3(S_i^y)^2]\}$$

$$M_{5,9} = M_{23}^y = -\bar{m}\{(_2C_m^y)^{-1}[(104L^2(S_i^y)^2 + 152k_{11}^y L^2 S_i^y)(S_j^y)^2$$
$$+ (260k_{11}^y L^2(S_i^y)^2 + 342(k_{11}^y)^2 L^2 S_i^y)S_j^y + 132(k_{11}^y)^2 L^2(S_i^y)^2$$
$$+ 147(k_{11}^y)^3 L^2 S_i^y]\}$$

$$M_{5,11} = M_{24}^y = \bar{m}\{(_2C_m^y)^{-1}[-(24L^3(S_i^y)^2 + 40k_{11}^y L^3 S_i^y)(S_j^y)^2$$
$$+ (40k_{11}^y L^3(S_i^y)^2 + 62(k_{11}^y)^2 L^3 S_i^y)S_j^y$$

$$M_{6,6} = M_{22}^z = \bar{m}\{(_2C_m^z)^{-1}[32L^3(S_i^z)^2(S_j^z)^2 + 88k_{11}^z L^3(S_i^z)^2 S_j^z$$
$$+ 64(k_{11}^z)^2 L^3(S_i^z)^2]\}$$

$$M_{6,8} = M_{22}^z = \bar{m}\{(_2C_m^z)^{-1}[(104L^2(S_i^z)^2 + 152k_{11}^z L^2 S_i^z)(S_j^z)^2$$
$$+ (260k_{11}^z L^2(S_i^z)^2 + 342(k_{11}^z)^2 L^2 S_i^z)S_j^z + 132(k_{11}^z)^2 L^2(S_i^z)^2$$
$$+ 147(k_{11}^z)^3 L^2 S_i^z]\}$$

$$M_{6,12} = M_{22}^z = \bar{m}\{(_2C_m^z)^{-1}[-(24L^3(S_i^z)^2 + 40k_{11}^z L^3 S_i^z)(S_j^z)^2$$
$$+ (40k_{11}^z L^3(S_i^z)^2 + 62(k_{11}^z)^2 L^3 S_i^z)S_j^z$$

$$M_{7,7} = 140\bar{m}L/420$$

$$M_{8,8} = M_{33}^z = \bar{m}\{(_1C_m^z)^{-1}[(416L(S_i^z)^2 + 936k_{11}^z LS_i^z + 544(k_{11}^z)^2 L)(S_j^y)^2$$
$$+ (656k_{11}^z L(S_i^z)^2 + 1312(k_{11}^z)^2 LS_i^z + 672(k_{11}^z)^3 L)S_j^y$$
$$+ 264(k_{11}^z)^2 L(S_i^z)^2 + 462(k_{11}^z)^3 LS_i^z + 210(k_{11}^z)^4 L$$

$$M_{8,12} = M_{34}^z = \bar{m}\{(_2C_m^z)^{-1}[-(176L^2(S_i^z)^2 + 440k_{11}^z L^2 S_i^z$$
$$+ 288(k_{11}^z)^2 L^2)(S_j^z)^2 + (128k_{11}^z L^2(S_i^z)^2 + 288(k_{11}^z)^2 L^2 S_i^z$$
$$+ 168(k_{11}^z)^3 L^2) S_j^z$$

$$M_{9,9} = M_{33}^y = \bar{m}\{(_1C_m^y)^{-1}[(416L(S_i^y)^2 + 936k_{11}^y L S_i^y + 544(k_{11}^y)^2 L)(S_j^y)^2$$
$$+ (656k_{11}^y L(S_i^y)^2 + 1312(k_{11}^y)^2 L S_i^y + 672(k_{11}^y)^3 L) S_j^y$$
$$+ 264(k_{11}^y)^2 L(S_i^y)^2 + 462(k_{11}^y)^3 L S_i^y + 210(k_{11}^y)^4 L$$

$$M_{9,11} = M_{34}^y = -\bar{m}\{(_2C_m^y)^{-1}[-(176L^2(S_i^y)^2 + 440k_{11}^y L^2 S_i^y$$
$$+ 288(k_{11}^y)^2 L^2)(S_j^y)^2 + (128k_{11}^y L^2(S_i^y)^2 + 288(k_{11}^y)^2 L^2 S_i^y$$
$$+ 168(k_{11}^y)^3 L^2) S_j^y$$

$$M_{10,10} = 140\bar{m}LJ/(420A)$$

$$M_{11,11} = M_{44}^y = \bar{m}\{(_2C_m^y)^{-1}[(32L^3(S_i^y)^2 + 88k_{11}^y L^3 S_i^y + 64(k_{11}^y)^2 L^3)(S_j^y)^2]\}$$

$$M_{12,12} = M_{44}^z = \bar{m}\{(_2C_m^z)^{-1}[(32L^3(S_i^z)^2 + 88k_{11}^z L^3 S_i^z + 64(k_{11}^z)^2 L^3)(S_j^z)^2]\}$$

where

$$k_{11}^z = 4EI_z/L$$
$$k_{11}^z = 4EI_y/L$$

$$(_1C_m^z)^{-1} = 2(560(S_i^z)^2 + 1120k_{11}^z S_i^z + 560(k_{11}^z)^2)(S_j^z)^2$$
$$+ (1120k_{11}^z(S_i^z)^2 + 1960(k_{11}^z)^2 S_i^z + 840(k_{11}^z)^3) S_j^z$$
$$+ 560(k_{11}^z)^2(S_i^z)^2 + 840(k_{11}^z)^3 S_i^z + 315(k_{11}^z)^4$$

$$(_2C_m^z)^{-1} = 2(1680(S_i^z)^2 + 3360k_{11}^z S_i^z + 1680(k_{11}^z)^2)(S_j^z)^2$$
$$+ (3360k_{11}^z(S_i^z)^2 + 5880(k_{11}^z)^2 S_i^z + 2520(k_{11}^z)^3) S_j^y$$
$$+ 1680(k_{11}^z)^2(S_i^z)^2 + 2520(k_{11}^z)^3 S_i^z + 945(k_{11}^z)^4$$

$$(_1C_m^z)^{-1} = 2(560(S_i^y)^2 + 1120k_{11}^y S_i^y + 560(k_{11}^y)^2)(S_j^y)^2$$
$$+ (1120k_{11}^y(S_i^y)^2 + 1960(k_{11}^y)^2 S_i^y + 840(k_{11}^y)^3) S_j^y$$
$$+ 560(k_{11}^y)^2(S_i^y)^2 + 840(k_{11}^y)^3 S_i^y + 315(k_{11}^y)^4$$

$$(_2C_m^z)^{-1} = 2(1680(S_i^y)^2 + 3360k_{11}^y S_i^y + 1680(k_{11}^y)^2)(S_i^y)^2$$
$$+ (3360k_{11}^y(S_i^y)^2 + 5880(k_{11}^y)^2 S_i^y + 2520(k_{11}^y)^3) S_j^y$$
$$+ 1680(k_{11}^y)^2(S_i^y)^2 + 2520(k_{11}^y)^3 S_i^y + 945(k_{11}^y)^4$$

2 TECHNIQUES IN TRANSIENT ANALYSIS OF SEMI-RIGID STEEL FRAMES

PETER PUI-TAK CHUI and SIU-LAI CHAN

Department of Civil and Structural Engineering,
The Hong Kong Polytechnic University, Hong Kong

2.1. SUMMARY

An efficient and robust large deflection dynamic analysis of steel framed structures with flexible and hysteretic beam-to-column connections is presented in this chapter. The objective is to predict a more realistic dynamic behaviour of steel framed structures accounting for nonlinear beam-to-column connection stiffness using a personal computer. The displacement-based finite element method (FEM) with the incremental-iterative procedure of equation of motion for structures undergoing arbitrarily large displacement but small strain is adopted. These equations are based on the updated Lagrangian description of coordinate system and are integrated with time by the Newmark integration numerical method. A reliable and simple hybrid element formed by a beam/column element with pseudo-rotational-springs at two ends accounting for connection nonlinear behaviour is used. The hysteretic loop characteristic of a semi-rigid connection is based on a known monotonic moment-rotation curve obtained from experiments. Geometric nonlinearity is also included in the present study and the so-called P-Delta effect due to axial force is studied through several numerical examples. The nonlinear behaviour of connections was found to affect structural dynamic response significantly because of the hysteretic damping effect leading to a dissipation of excitation energy at connections. It is interesting to note that the hysteretic damping cannot be exhibited in the rigid, pinned nor linear connections. The influence of flexible connections on overall structural response will be studied and discussed through the numerical examples in this chapter.

2.2. INTRODUCTION

In the conventional analysis and design of steel structures, beam-column joints are usually assumed to be either perfectly rigid or ideally pinned for simplicity. This implies that the angle between the adjoining members remains unchanged for the rigid joint case and no moment will be transferred between the adjoining members for pinned joint case. However, in reality, experiments (Popov and Stephen, 1970; Popov, 1983; Nader and Astaneh, 1991) demonstrated that connections behave nonlinearly in a manner between the two extreme cases. This means that a finite degree of joint flexibility existed at connections. Connections in practice are, therefore, semi-rigid. The influence of semi-rigid connections on realistic structural response has been recognized and provision for semi-rigid connections has been given in some national steel design codes such as the American Load and Resistance Factored Design (LRFD) (1986), the British Standard BS5950 (1985), and the Eurocode 3 (1988). For example, the LRFD code (1986) proposed by the American Institute of Steel Construction (AISC) permits two basic types of constructions: the Type FR (fully restrained) construction which is commonly designated as rigid frame, and the Type PR (partially restrained) construction which allows for semi-rigid connection stiffness. Although the LRFD code permits a partial restraint to be utilized, it gives little guidance to engineers on the utilization of finite connection stiffness.

Experimental results showed that the practical connections behave non-linearly due to gradual local yielding of connection plates and cleats, etc. The flexible connections in low-rise to medium-rise moment-resisting frames affect significantly the overall structural performance such as deformations, force distributions and dynamic response. In general, a flexibly connected steel frame experiences a larger lateral drift but may attract lesser storey shear forces when compared with a rigid jointed frame. Further, this nonlinear behaviour of connections in a ductile steel frame will exhibit non-recoverable hysteretic damping which is a primary source of passive damping during the vibration of a structure. This damping cannot be observed from a frame with rigid, pinned or linear (constant stiffness) connection types. The amount of energy dissipated in a connection in each cycle response is equal to the area enclosed in the moment-rotation hysteretic loop.

In the dynamic analysis, there are three primary sources of dampings. They are the viscous damping, the hysteretic damping at nonlinear connections and the hysteretic damping at plastic hinges. In order to focus attention on the influence of hysteretic damping at connections to the overall structural response, the material, however, is assumed to be elastic throughout the analysis in this chapter. Similarly, the viscous damping is also ignored.

Generally speaking, a building can survive during an earthquake ground excitation if its energy absorption capacity is larger than the input energy. It implies the advantage of using semi-rigid jointed structures in a seismic zone. In this case, the semi-rigid connection reduces the structural stiffness and hence the structure will exhibit a larger drift which is, although, not desired in serviceability limit state but may be acceptable in seismic design of structures. The hysteretic damping at connections will increase the energy-absorption capacity of a structure. Moreover, the storey shear forces may be reduced because of lesser inertial forces induced in a softer structure. In other words, the seismic response may be stabilizing for semi-rigid frames. Thus, the dynamic response of flexibly connected steel structures is considerably affected by the connections when compared with the static analysis. In summary, connection flexibility should be considered in the dynamic analysis of steel frames.

Unlike brittle materials such as concrete, steel has a high ductility and thus a stronger resistance to cyclic and repeated loads. Besides, if a connection is properly detailed, its connection properties including joint stiffness can be retained and hence the hysteretic moment-rotation $M - \phi$ loops occur under repeated loading and unloading conditions. Furthermore, experiments (Popov and Pinkney, 1969; Vedero and Popov, 1972; Tsai and Popov, 1990; Azizinamini and Radziminski, 1989) demonstrated that the hysteretic loop at connections is stable and reproducible under cyclic loads. Therefore the static monotonic moment-rotation curves can be employed and extended to model the behaviour of connection hysteretic loops in the present dynamic analysis.

Due to the intensive and extensive effort and cost for conducting an experiment, numerical experiments are generally preferred for large-scale structures. Based on tested moment-rotation curves, an efficient and robust three-dimensional nonlinear geometric large deflection dynamic analysis of steel structures accounting for nonlinear and hysteretic beam-to-column connections is described in this chapter. The nonlinear behaviour of connections was found to affect significantly the overall structural performance and should be included in the dynamic analysis for accurate prediction of the dynamic response of steel buildings.

2.3. REVIEW OF EXPERIMENTAL AND ANALYTICAL RESEARCH IN CONNECTIONS

2.3.1. Review of Monotonic Static Loading Tests

Since the possibly earliest studies of rotational stiffness of the beam-column connections by Wilson and Moore (1917), hundreds of tests have been conducted to establish the relationship between moments and relative rotations of beam-column connections. Prior to 1950, riveted connections were tested by Young and Jackson (1934) and Rathbun (1936). In parallel with the interest in using high-strength bolts as structural fasteners, this connection type was tested by Bell *et al.* (1958). Subsequently, the behaviour of header plate connection was investigated in 20 tests by Sommer (1969).

The extended end-plate and the flush end-plate connections have been used extensively since the late 1960's. The extended end-plate connections are designed to transfer considerable moments from beams to columns. Flush end-plate connections and extended end-plate connections were tested by Ostrander (1970) and Johnstone and Walpole (1981) respectively.

Davison *et al.* (1987) performed a series of tests on a variety of beam-to-column connections including the web-cleat, flange cleat, combined seating cleat and web cleats, flush end-plate and extended end-plate connections. An informative data of connection moment-rotation $M - \phi$ curves was collected. More recently, Moore *et al.* (1993a,b) carried out tests on five full-scale steel frames. The connections tested are a combination of flush end-plate connection, extended end-plate connection and flange cleat connection. Complete records of deformation of members and joints were reported. The results were also compared with the current British design code BS5950 Part 1, and extensions and modification to the existing code have been proposed for better prediction of analysis results accounting for the role of the joints.

Simultaneously, a number of mathematical models were proposed to represent the moment versus rotation curves of the tested connections. In 1975, Frye and Morris (1975) collected a total of 145 test results in the period from 1936 to 1970. Seven types of connection test data were classified, covering the range from the weakest single web angle connection to the stiffest T-stud connection. They also modelled the moment-rotation relationship by a polynomial function which was previously adopted by Sommer (1969). Later, Ang and Morris (1984) further collected 32 experimental results from 1934 to 1976, which were grouped into five types including the strap angle connection. The more accurate Ramberg-Osgood function (Ramberg and Osgood, 1943) was adopted to represent the standardized moment-rotation relationship for the five commonly used connection types. Besides, Goverdhan (1984) gathered a total of 230 test results of moment-rotation curves which were digitized to form a database of connection behaviour. In 1985, Nethercot (1985) conducted a literature survey for the period from 1951 to 1985. The connection test data and the corresponding curve representations were reviewed. Kishi and Chen (1986a; 1986b; 1989) further extended the collection to include over 300 test results which are classified into seven connection types. Namely, they are the single/double web-angle, top- and seat-angle with/without double web-angle, extended/flush end-plate and header plate connections. A computerized data bank program called the Steel Connection Data Bank (SCDB) was developed on steel beam-to-column connections to monitor the data base systematically. Using the system of tabulation and plotting in the SCDB program, the moment-rotation characteristics and the appropriate analytical connection model for the seven commonly used connection types can be easily obtained and used in analysis.

2.3.2. Review of Cyclic Loading Tests

Unlike the static monotonic tests of connection behaviour, only relatively limited number of connections were tested under cyclic and repeated loading in the past few years. In late 1960's and early 1970's, the cyclic behaviour of connections was studied on small specimens under severe loads by Popov and Pinkney (1969). Subsequently, more comprehensive studies of connection behaviour as parts of the one-third scale interior column sub-assemblages were reported by Krawinkler et al. (1971) and Carpenter and Lu (1973). Later, Clough and Tang (1975) tested the moment resisting frames with realistic joints on scaled models on a shaking table. In order to obtain more realistic results of connection behaviour, full-scale experiments on welded beam to column moment connections and bolted web and welded flanges connections were conducted by Popov and his co-workers (Popov and Stephen, 1970; Popov and Bertero, 1973; Popov, 1983).

Recently, Stelmack et al. (1986) conducted ten tests on full scale steel frames with top and seat flange angle connections under cyclic loads. In order to give a more reliable result of moment-rotation curves, a total of 14 specimens of 1/2 inch thick angle connections were tested under monotonic loads. The experimental response of load-deflection curves and moment-rotation curves were obtained. A trilinear function for moment-rotation $M - \phi$ curves was also proposed to model the angle connection and comparison between the analytical and the experimental results was reported.

Subsequently, the hysteretic behaviour of top and seat flange angles with double web angle connection for various geometric parameters such as connection thickness, bolt size and spacing were investigated under constant and variable amplitude cyclic loads by Azizinamini and Radziminski (1989).

Portal frames with similar connection type (top and seat bolted flange with double bolted web angle connection) as well as the pinned (double bolted angle web) and rigid (welded flange and angle bolted web) connection types were analysed by Nader and Astaneh (1991). The frames were subjected to various earthquake attacks using the shaking table. A total of 44 shaking table tests were carried out and it was observed that the semi-rigid frames reduced the base shear but the lateral drift was increased under ground excitation when compared to the rigid frames.

In usual practice, the shop-welded and site-bolted procedure is generally adopted. The behaviour of the commonly used extended end-plate connections under cyclic loading was reported by Tsai et al. (1990) and Korol et al. (1990). Tsai et al. (1990) performed three cyclic tests on a large-size beam with extended end-plate connections and it was observed that the specimen with larger connection bolts and a slightly thicker end plate had a superior ductility and it is recommended for design of structures against seismic loads. Korol et al. (1990) carried out tests on seven full-scale extended end plate connections under cyclic loading and concluded that sufficient energy dissipation capability without substantial loss of strength can be achieved by proper detailing of the connection.

2.3.3. Review of Analysis Methods

In static analysis, the methods used for analyzing steel frames with flexible connections include the linear analysis, the bifurcation analysis (Cosenza et al., 1989) and its associated method for ultimate load determination based on the Merchant-Rankise formula (Jaspart, 1988), the semi-bifurcation analysis (Ho and Chan, 1991), the inelastic analysis (Al-Bermani and Kitipornchai, 1992; Chen and Zhou, 1987; Savard et al., 1994), the second order nonlinear analysis using the total and the incremental secant stiffness matrix (Goto and Chen, 1987; Anderson and Benterkia, 1991; Ho and Chan, 1993) and the ultimate analysis (Lui and Chen, 1987; Chan and Yau, 1992).

Although the static analysis of flexibly connected frames has been carried out extensively, research on dynamic analysis has received relatively little attention. Kawashima and Fujimoto (1984) conducted the experimental investigation on the vibration of steel frames with semi-rigid connections. The connection stiffness was assumed to remain constant in their analysis and the tested natural frequency was compared with the analytical solutions. However, their transient response was not investigated. The seismic behaviour of two realistic 10 and 20 storey all-steel moment-resistant office buildings located in Canada were studied by Sivakumaran (1988). The connection stiffness was modelled by a bilinear moment-rotation function which is only acceptable for small loading range. Numerical procedure to handle the connection flexibility was not discussed in his paper.

Using a developed computer package program, DRAIN-2D, a single-bay two-storey unbraced frames with top and seat angle connections under cyclic load was investigated by Youssef-Agha *et al.* (1989). Based on the geometric properties and configuration of a connection, the moment-rotation curve was represented by a bilinear function. However, only the modelling of the top-and seat angle connection type was given and the model is rather coarse.

Recently, Shi and Atluri (1989) proposed a numerical method based on the complementary energy principle to study the dynamic and large deflection response of semi-rigid steel frames. Gao and Haldar (1995) developed a numerical method using the assumed stress-based finite-element method with the tangent stiffness formulation. In their work, the generally accepted displacement-based finite element procedure was abandoned. In the present study, a robust and efficient method based on the conventional beam-column element for the large deflection elastic analysis allowing the flexibility of beam-to-column connections in steel frames under dynamic loads is presented. The suggested method uses on the displacement-based finite element numerical procedure which is more widely known and familiar to the engineers and researchers.

2.4. BEHAVIOUR AND MODELLING OF CONNECTIONS

It is well known that the widely used assumption of perfectly rigid and frictionless pinned beam-column connection is practically unattainable. Numerous experiments have shown that practical connections behave non-linearly due to gradual yielding of connection plate, bolts etc. This nonlinear behaviour of a connection can be represented by the moment-rotation curve which is generally obtained from test results. Typical moment-rotation curves of several commonly used connections are shown in Figure 1. In order to incorporate a moment-rotation curve more systematically and efficiently into a frame analysis, the moment versus rotation relationship is usually modelled by a mathematical function. A good mathematical function should be simple, requiring fewer parameters, easy in determination of these parameters, physically meaningful, stable, containing no negative derivative, and capable of representing a wide range of connection types.

In reality, the nonlinear behaviour of connections plays an important role in structural response, especially in dynamic case. Under dynamic loads, hysteretic loops of moment-rotation curve in connections will be resulted

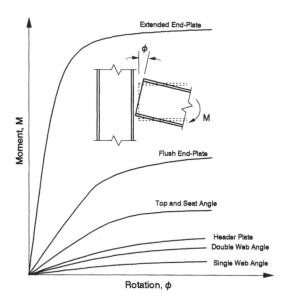

Figure 1. Typical moment-rotation curves of beam-to-column connections.

and the passive hysteretic connection damping is generated in the structure. In other words, the nonlinear beam-column connections in a steel building possess a capability of dissipating excitation energy by an amount equal to the enclosed loop area under cyclic loads. This hysteretic damping characteristic at connections cannot be utilized or made use of in the traditional design for rigid frames and pinned joint trusses and its effect is particularly important for moment resistant buildings against seismic attack.

Tests on typical beam-column connections showed that the moment versus rotation curves are stable and reproducible (Popov and Pinkney, 1969; Vedero and Popov, 1972; Tsai and Popov, 1990; Azizinamini and Radziminski, 1989). If connections are properly detailed, they will have a highly ductile behaviour on moment-rotation loops (Korol et al., 1990). Based on the above experimental observation, studies in this chapter assume that the static monotonic moment-rotation curve can be used and extended to dynamic analysis. On this basis, stable and reproducible hysteretic loops of moment-rotation curve in a connection can be generated, as shown in Figure 2.

There are several types of mathematical models for simulation and representation of the moment-rotation curves. Namely, they are the linear

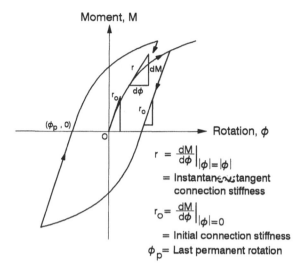

$$r = \frac{dM}{d\phi}\bigg|_{|\phi|=|\phi|}$$

= Instantaneous tangent
connection stiffness

$$r_o = \frac{dM}{d\phi}\bigg|_{|\phi|=0}$$

= Initial connection stiffness

ϕ_p = Last permanent rotation

Figure 2. A typical hysteretic loop of moment-rotation $M - \phi$ curve of a nonlinear flexible connection.

model (Arbabi, 1982; Kawashima and Fujimoto, 1984), the bilinear model (Sivakumaran, 1988; Youssef-Agha, 1989), the trilinear model (Stelmack *et al.*, 1986; Gerstle, 1988), the polynomial model (Frye and Morris, 1975), the cubic B-Spline model (Cox, 1972; Jones *et al.*, 1980), the Ramberg-Osgood model (Ramberg and Osgood, 1943; Shi and Atluri, 1989), the Richard-Abbott model (Richard and Abbott, 1975; Gao and Haldar, 1995), the exponential model (Lui and Chen, 1988), and among others. The linear and the bilinear models are simple to use but they may be too coarse and produce drastic change in stiffness which is undesirable in terms of accuracy and computational stability. The polynomial model is able to provide a better approximation, but could give undesired negative connection stiffness. The cubic B-Spline model can represent the results accurately but requires many parameters. The Ramberg-Osgood and the Richard-Abbott models require respectively three and four parameters and give a fairly good fit. The exponential model can provide an excellent fit and at least six parameters are used in modeling. Further, the exponential, the Ramberg-Osgood and the Richard-Abbott models can always give a positive derivative which is physically more acceptable. Thus, these three models are adopted in modelling of connection in the analysis of frames in this chapter.

Under cyclic loads, the hysteretic loop as well as the instantaneous tangent connection stiffness on the moment-rotation ($M - \phi$) curve will change in accordance with the current value of moment M, the relative slip rotation ϕ and the increment/decrement of moment ΔM at a connection. In order to describe the behaviour of a connection under cyclic loading, the following algorithm is used to monitor the connection stiffness in a computer analysis (see also Figure 2 for ease of reference).

(1) If the path starts from or passes through a point (ϕ_p, 0) originated on the line $M = 0 (M \cdot \Delta M > 0)$, it will follow the curve of the mathematical model by taking the point (ϕ_p, 0) as the new or updated origin.

(2) If the path is undergoing unloading ($M \cdot \Delta M < 0$), it will move reversely along the line of slope of the initial stiffness r_0 and this absolute maximum moment $|M|$ will be stored for later use.

(3) If the connection is re-loaded ($M \cdot \Delta M > 0$) before reaching the line $M = 0$, it will follow a line parallel to the initial stiffness r_0 until it reaches the maximum moment $|M|$ previously stored. When it is further loaded ($M \cdot \Delta M > 0$), it will follow the curve originated at the previous permanent rotation point (ϕ_p, 0).

The algorithm described above is applicable to any connection models, so many existing valuable laboratory test data (Nethercot, 1985; Kishi and Chen, 1986a and b; Lui and Chen, 1988) can be employed and incorporated into the computer code directly without an expensive and time-consuming curve-fitting procedure from one model to another. The basic characteristics of the three models for a nonlinear connection under cyclic loads are summarized as follows.

The Exponential Model

In the exponential model, the virgin moment-rotation curve under loading condition is expressed by,

$$M = \sum_{j=1}^{n} C_j \left[1 - \exp\left(\frac{-|\phi|}{2j\alpha} \right) \right] + R_{kf} |\phi| + M_0 \quad \text{when } M \cdot \Delta M > 0 \quad (1)$$

and its tangent connection stiffness will be given by,

$$r = \left. \frac{dM}{d\phi} \right|_{|\phi|=|\phi|} = \sum_{j=1}^{n} \frac{C_j}{2j\alpha} \left[\exp\left(\frac{-|\phi|}{2j\alpha} \right) \right] + R_{kf} \quad \text{when } M \cdot \Delta M > 0 \quad (2)$$

in which M is the moment in the connection, ϕ is the absolute value of the rotation deformation of the connection, M_0 is the initial moment, R_{kf} is the strain-hardening stiffness of the connection, and C_j and α are the curve-fit coefficients. When the path is undergoing unloading or reloading, the stiffness will be taken as the initial stiffness as,

$$r_0 = \frac{dM}{d\phi}\bigg|_{|\phi|=0} = \sum_{j=1}^{n} \frac{C_j}{2j\alpha} + R_{kf} \text{ when } M \cdot \Delta M \cdot < 0 \qquad (3)$$

The Ramberg-Osgood Model

The loading moment-rotation curve in this model can be expressed as,

$$\frac{\phi}{\phi_0} = \frac{|KM|}{(KM)_0}\left[1 + \left(\frac{|KM|}{(KM)_0}\right)^{n-1}\right] \quad \text{when } M \cdot \Delta M > 0 \qquad (4)$$

with tangent connection stiffness as,

$$r = \frac{(KM)_0/\phi_0}{1 + n\left[\frac{|KM|}{(KM)_0}\right]^{n-1}} \quad \text{when } M \cdot \Delta M > 0 \qquad (5)$$

in which $(KM)_0$ is a reference moment, ϕ_0 is a reference rotation, n is a parameter defining the sharpness of the curve and K is a standardization constant dependent on the connection type and geometry. Under unloading and reloading conditions, the connection stiffness can be written as,

$$r_0 = \frac{dM}{d\phi}\bigg|_{M=0} = \frac{(KM)_0}{\phi_0} \quad \text{when } M \cdot \Delta M < 0 \qquad (6)$$

The influence of the parameter, n, to the shape of the Ramberg-Osgood model is shown in Figure 3(a)(b). It can been seen that the ultimate moment will be more flatten with increasing the value of the parameter, n.

The Richard-Abbott Model

In the virgin loading path, the moment-rotation behaviour is given by the following expression in this model as,

$$M = \frac{(k - k_p)|\phi|}{\left[1 + \left|\frac{(k-k_p)|\phi|}{M_0}\right|^n\right]^{1/n}} + k_p|\phi| \quad \text{when } M \cdot \Delta M > 0 \qquad (7)$$

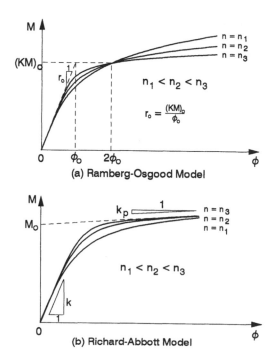

Figure 3. Forms of models.

and the corresponding tangent stiffness by

$$r = \left. \frac{dM}{d\phi} \right|_{|\phi|=|\phi|} = \frac{(k - k_p)}{\left[1 + \left| \frac{(k-k_p)|\phi|}{M_0} \right|^n \right]^{n+1/n}} + k_p \quad \text{when } M \cdot \Delta M > 0 \quad (8)$$

in which k is the initial stiffness, k_p is the strain-hardening stiffness, M_0 is a reference moment and n is a parameter which defines the sharpness of the curve. The family of plots of the Richard-Abbott model is depicted in Figure 3(b), corresponding to three values of n. For the unloading or reloading case, the connection stiffness is simply given by,

$$r_0 = k \qquad \text{when } M \cdot \Delta M < 0 \qquad (9)$$

Using the algorithm described and equations (1) to (9), reproducible hysteretic loops at a connection can be described.

2.5. NUMERICAL PROCEDURE FOR NONLINEAR TRANSIENT ANALYSIS

The following section is to introduce the numerical procedure accounting for the connection flexibility in transient analysis performed in the present study. Before this, the following assumptions are made.

2.5.1. Assumptions

The assumptions made in this chapter are as follows:

(1) The Bernoullis' assumption that a plane section normal to the centroidal axis before deformation remains plane after deformation and normal to the axis is made. Warping shear and cross-section distortion are not considered.
(2) Small strains but arbitrarily large displacements and rotations are considered.
(3) The material is assumed to remain elastic throughout the whole loading range. Nonlinearity is due to joint flexibility and geometrical change.
(4) All connections are assumed to be detailed properly and retain their connection stiffness and characteristic under cyclic loading.
(5) The hysteretic loop at a connection is based on its static monotonic moment-rotation curve which is fitted by a mathematical function.
(6) The connection element is of zero length.

2.5.2. Numerical Solution Method

In the present study, the generally accepted displacement-based finite element method is used. The equations of motion are based on the coordinate system of the updated Lagrangian description and are expressed as an incremental-iterative procedure. In the time-domained dynamic analysis, the time history is discreted into a number of equally spaced time steps. Based on the last known equilibrium solution, the solution at the next time step is solved successively, starting from the initial condition until to the desired time of study. The initial condition here is assumed to be stationary before the time starts. By a step to step procedure, the unconditionally stable Newmark numerical integration method with the constant-average-acceleration assumption is employed (Newmark, 1950; Bathe, 1996). Iterations are performed for each time step to satisfy equilibrium before marching to the next step. Within a time step, tangent connection stiffness is assumed constant in order to eliminate the possible connection stiffness oscillation under the iterative loading/unloading process. Due to the geometrical change, the tangent stiffness can be reformed and updated in each iteration in order to accelerate the iterative procedure within a time step. However, it require a longer computational time in reforming the updated tangent stiffness. Alternatively the constant stiffness iteration scheme, in which the tangent stiffness is kept constant, can be chosen. Although it does not require such a tangent stiffness reforming procedure, it needs more number of iterations. Therefore, a combination of them (i.e. the tangent stiffness re-built for several iterations) is considered.

1.5.3. Formulation of Linear Stiffness Matrix

In the present study, a hybrid element with two end springs of zero length is employed because of its simplicity and reliability tested by Chen and Chan (1994) and Yau and Chan (1994) in their static analysis. This element will be extended to the dynamic problems in the present analysis.

A detailed description of formulation for the stiffness matrix of the element has been given by Yau and Chan (1994) and will be recapitulated here for clarity. Due to the flexibility of a semi-rigid joint, the rotational displacements on the two sides of a connection are the internal and the external rotations (i.e. $_i\theta$ and $_e\theta$ respectively) which are generally unequal (see Figure 4). The subscripts of "i" and "e" refer to the internal and external nodes relative to the element.

Considering the moment equilibrium conditions at a connection as shown in Figure 4, the following stiffness equation can be written as,

$$\begin{bmatrix} \Delta_e M \\ \Delta_i M \end{bmatrix} = \begin{bmatrix} r & -r \\ -r & r \end{bmatrix} \begin{bmatrix} \Delta_e \theta \\ \Delta_i \theta \end{bmatrix} \tag{10}$$

in which $\Delta_e M$ and $\Delta_i M$ are the incremental nodal moments at the junctions between the spring and the global node and between the beam and the spring, $\Delta_e \theta$ and $\Delta_i \theta$ are the incremental nodal rotations corresponding to these moments and r is the instantaneous connection stiffness given by,

$$r = \frac{dM}{d\phi} \tag{11}$$

in which M is the moment at the joint and ϕ is the slip angle defined by,

$$\phi = {}_e\theta - {}_i\theta \tag{12}$$

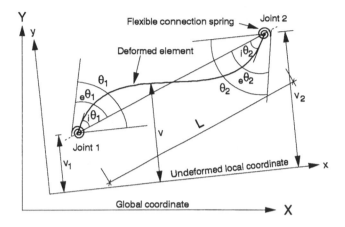

Figure 4. The lateral deflections and rotations of a deformed element with flexible connection springs at the ends.

After combining the spring stiffness to the two ends of an element and expressing in incremental form, we have,

$$\begin{bmatrix} \Delta_e M_1 \\ \Delta_i M_1 \\ \Delta_i M_2 \\ \Delta_e M_2 \end{bmatrix} = \begin{bmatrix} r_1 & -r_1 & 0 & 0 \\ -r_1 & r_1 + \frac{4EI}{L} & \frac{2EI}{L} & 0 \\ 0 & \frac{2EI}{L} & r_2 + \frac{4EI}{L} & -r_2 \\ 0 & 0 & -r_2 & r_2 \end{bmatrix} \begin{bmatrix} \Delta_e \theta_1 \\ \Delta_i \theta_1 \\ \Delta_i \theta_2 \\ \Delta_e \theta_2 \end{bmatrix} \tag{13}$$

in which the subscript "1" and "2" are referred to the node 1 and node 2. Assuming the loads are applied only at the global nodes and hence both $\Delta_e M_1$ and $\Delta_e M_2$ are equal to zero, we obtain,

$$\begin{bmatrix} \Delta_i \theta_1 \\ \Delta_i \theta_2 \end{bmatrix} = \begin{bmatrix} r_1 + \frac{4EI}{L} & \frac{2EI}{L} \\ \frac{2EI}{L} & r_2 + \frac{4EI}{L} \end{bmatrix}^{-1} \begin{bmatrix} r_1 & 0 \\ 0 & r_2 \end{bmatrix} \begin{bmatrix} \Delta_e \theta_1 \\ \Delta_e \theta_2 \end{bmatrix} \tag{14}$$

Eliminating the internal degrees of freedom by substituting the Equation (14) into (13), we have,

$$\begin{bmatrix} \Delta_e M_1 \\ \Delta_e M_2 \end{bmatrix} = \left(\begin{bmatrix} r_1 & 0 \\ 0 & r_2 \end{bmatrix} - \frac{\begin{bmatrix} r_1 & 0 \\ 0 & r_2 \end{bmatrix} \begin{bmatrix} r_2 + \frac{4EI}{L} & -\frac{2EI}{L} \\ -\frac{2EI}{L} & r_1 + \frac{4EI}{L} \end{bmatrix} \begin{bmatrix} r_1 & 0 \\ 0 & r_2 \end{bmatrix}}{\left(r_1 + \frac{4EI}{L} \right) \left(r_2 + \frac{4EI}{L} \right) - 4 \left(\frac{EI}{L} \right)^2} \right) \begin{bmatrix} \Delta_e \theta_1 \\ \Delta_e \theta_2 \end{bmatrix} \tag{15}$$

Transforming the incremental moments, $\Delta_e M_1$ and $\Delta_e M_2$, in Equation (15) to the moments and shears about the last known axes, we have,

$$
\begin{bmatrix} \Delta M_1 \\ \Delta Q_1 \\ \Delta M_2 \\ \Delta Q_2 \end{bmatrix} = \begin{bmatrix} 1 & 0 \\ \frac{1}{L} & \frac{1}{L} \\ 0 & 1 \\ -\frac{1}{L} & -\frac{1}{L} \end{bmatrix} \times
$$

$$
\left(\begin{bmatrix} r_1 & 0 \\ 0 & r_2 \end{bmatrix} - \frac{\begin{bmatrix} r_1 & 0 \\ 0 & r_2 \end{bmatrix} \begin{bmatrix} r_2 + \frac{4EI}{L} & -\frac{2EI}{L} \\ -\frac{2EI}{L} & r_1 + \frac{4EI}{L} \end{bmatrix} \begin{bmatrix} r_1 & 0 \\ 0 & r_2 \end{bmatrix}}{\left(r_1 + \frac{4EI}{L} \right) \left(r_2 + \frac{4EI}{L} \right) - 4 \left(\frac{EI}{L} \right)^2} \right) \tag{16}
$$

$$
\begin{bmatrix} 1 & \frac{1}{L} & 0 & -\frac{1}{L} \\ 0 & \frac{1}{L} & 1 & -\frac{1}{L} \end{bmatrix} \begin{bmatrix} \Delta\theta_1 \\ \Delta v_1 \\ \Delta\theta_2 \\ \Delta v_2 \end{bmatrix}
$$

in which ΔM_1 and ΔM_2 are the incremental moments and ΔQ_1 and ΔQ_2 are the incremental shears at the two ended nodes, r_1 and r_2 are the tangent spring stiffness at the ends of the element, $\Delta\theta_1$ and $\Delta\theta_2$ are the incremental rotations at the two ends of an element about the axis parallel to the last axis, Δv_1 and Δv_2 are the incremental lateral displacements at the two nodes of the element and projected onto the last configuration of the element, and L is the element length (see Figure 4).

The axial and the torsional stiffness given by EA/L and GJ/L will be superimposed to the above bending stiffness to form a complete linear stiffness matrix of the element.

1.5.4. Formulation of Consistent Mass Matrix

In the present studies, the consistent mass matrix based on the deflected shape allowing for the semi-rigid connection is derived. In most previous studies, the lumped mass matrix was used for simplification of calculation by ignoring the rotational masses. The consistent mass matrix is based on the shape function of deflection of the element which is derived in the following.

Considering the geometrical compatibility of the boundary condition at the ends of a beam-column element (see Chan and Ho, 1994 for details), the commonly used cubic Hermitian function for the lateral-deflection of an arbitrary point along the deflected curve is the sum of the deflections due to the nodal rotations and the translational nodal displacements. It can be expressed in terms of the ended internal rotations, $_i\theta_1$ and $_i\theta_2$, and nodal displacements, v_1 and v_2 (see Figure 4), as

$$v = [\rho_1^2\rho_2L - \rho_1\rho_2^2L][_i\theta_1 \ _i\theta_2]^T + \rho_1v_1 + \rho_2v_2 \qquad (17)$$

in which $\rho_1 = 1 - x/L$, $\rho_2 = x/L$ and L is the length of the element.

Substituting Equation (14) into (17) and transforming the external nodal rotations, $_e\theta_1$ and $_e\theta_2$, about the convected axis along the ends of the element to the global nodal rotations, θ_1 and θ_2, the final displacement function, v, can be written as,

$$v = [\rho_1^2\rho_2L - \rho_1\rho_2^2L]\begin{bmatrix} \dfrac{4EI}{L}+r_1 & \dfrac{2EI}{L} \\ \dfrac{2EI}{L} & \dfrac{4EI}{L}+r_2 \end{bmatrix}^1 \begin{bmatrix} r_1 & 0 \\ 0 & r_2 \end{bmatrix} \qquad (18)$$

$$\begin{bmatrix} \dfrac{1}{L} & 1 & -\dfrac{1}{L} & 0 \\ \dfrac{1}{L} & 0 & -\dfrac{1}{L} & 1 \end{bmatrix} \begin{bmatrix} v_1 \\ \theta_1 \\ v_2 \\ \theta_2 \end{bmatrix} + \rho_1v_1 + \rho_2v_2$$

in which θ_1 and θ_2 are the nodal rotations of the element about the global axis.

Equation (18) represents the displacement function expressed in terms of the nodal displacements and the spring stiffness at two ends. It can be used directly in the formulation of the mass matrix as follows.

$$[M] = \int_0^L \bar{m}vv^T dx \qquad (19)$$

in which $\bar{m}$ is the mass per unit length of the element. The mass matrix [M] for prismatic members can be, therefore, evaluated directly and explicitly.

2.5.5. Formulation of Geometric Stiffness Matrix

In order to consider the instability effect of the element, the geometric stiffness matrix $[\mathbf{K_G}]$ is included in the incremental equilibrium equation by addition to the linear stiffness matrix $[\mathbf{K_L}]$ to form the instantaneous tangent stiffness matrix. Considering only the instability effect due to the axial load P in the element, we have,

$$[K_G] = \frac{P}{2} \int_0^L \left(\frac{\partial v}{\partial x}\right) \left(\frac{\partial v}{\partial x}\right)^T dx \tag{20}$$

Substituting Equation (18) into (20), the explicit form of the geometric stiffness matrix accounting for the connection spring stiffness can be evaluated directly.

2.5.6. Formulation of Viscous Damping Matrix

Similar to the consistent mass matrix, the viscous damping matrix may be written in the following form,

$$[C] = \int_0^L c(x) v v^T dx \tag{21}$$

where $c(x)$ is the distributed damping coefficient per unit length. However, in practice, the evaluation of the damping property $c(x)$ is complex and may be unreliable. For this reason, the viscous damping matrix is usually assumed to be proportional to the mass and stiffness matrices (Clough and Penzien, 1975) as,

$$[C] = a[M] + b([K_L] + [K_G]) \tag{22}$$

in which a and b are proportional constants that can be evaluated from the natural frequencies of the structure (Leger and Dussault, 1992). In order to demonstrate the effect of hysteretic damping at connections, the proportional damping is not considered here although it can be easily included in the analysis.

2.5.7. Equations of Motion

In transient analysis, the equilibrium equation at time $t + \Delta t$ is sought on the basis of the obtained equilibrium state at time t. The incremental equilibrium equation can be written as,

$$[M][\Delta \ddot{u}] + [C][\Delta \dot{u}] + ([K_L] + [K_G])[\Delta u] = [\Delta F] \tag{23}$$

in which $[M]$ and $[C]$ are the mass and the viscous damping matrices, $[K_L]$ and $[K_G]$ are the linear and the geometric stiffness matrices, $[\Delta F]$ is the incremental applied force vector, $[\Delta u]$, $[\Delta \dot{u}]$ and $[\Delta \ddot{u}]$ are the incremental displacement, velocity and acceleration vectors respectively.

In case of seismic loads, the incremental load vector $[\Delta F]$ can be expressed as,

$$[\Delta F] = -[M][E]\Delta \ddot{x}_g \tag{24}$$

in which $[E]$ is an index vector of the inertia forces and $\Delta \ddot{x}_g$ is an increment of absolute ground acceleration. In this case, $[\Delta u]$, $[\Delta \dot{u}]$ and $[\Delta \ddot{u}]$ are the incremental displacement, velocity and acceleration vectors relative to the ground respectively.

Expressing the velocity and the acceleration vectors at time $t + \Delta t$ in terms of the known displacement, velocity and acceleration vectors at time t and the acceleration vector at time $t + \Delta t$ by the Newmark method (Newmark, 1959), we have

$$[^{t+\Delta t}\dot{u}] + [^{t}\dot{u}] + (1 - \beta)\Delta t[^{t}\ddot{u}] + \beta \Delta t[^{t+\Delta t}\ddot{u}] \tag{25}$$

$$[^{t+\Delta t}\ddot{u}] + [^{t}u] + \Delta t[^{t}\dot{u}] + (\frac{1}{2} - \alpha)\Delta t^2[^{t}\ddot{u}] + \alpha \Delta t^2[^{t+\Delta t}\ddot{u}] \tag{26}$$

and hence the final incremental equilibrium equation for solving the unknown displacement increment $[\Delta u]$ at time $t + \Delta t$ can be expressed as,

$$[K]_{eff}[\Delta u] = [\Delta F]_{eff} \tag{27}$$

where

$$[K]_{eff} = [K_L] + [K_G] + \frac{1}{\alpha(\Delta t)^2}[M] + \frac{\beta}{\alpha \Delta t}[C] \tag{28}$$

$$[\Delta F]_{eff} = [\Delta F] + [M]\left(\frac{[^{t}\dot{u}]}{\alpha \Delta t} + \frac{[^{t}\ddot{u}]}{2\alpha}\right) + [C]\left(\frac{\beta}{\alpha}[^{t}\dot{u}] + \frac{\Delta t}{2}\left(\frac{\beta}{\alpha} - 2\right)[^{t}\ddot{u}]\right) \tag{29}$$

in which $[K]_{eff}$ and $[\Delta F]_{eff}$ are the effective tangent stiffness and the effective load increment, α and β are the Newmark's parameters taken as 0.25 and 0.5 for the constant-average-acceleration assumption within each time step, $[^{t}\dot{u}]$ and $[^{t}\ddot{u}]$ are the total velocity and acceleration at time t. It is noted that the form of equation (27) is similar to an equivalent static equation except inertial forces due to the mass and damping are included here.

Once the displacement increment $[\Delta u]$ at time $t + \Delta t$ in Equation (27) is solved, it is used to update the geometry of the structure, the displacement, the acceleration and the velocity by the following equations.

$$[^{t+\Delta t}x] = [^{t}x] + [\Delta u] \tag{30}$$

$$[^{t+\Delta t}u] = [^{t}u] + [\Delta u] \tag{31}$$

$$[^{t+\Delta t}\ddot{u}] = \frac{1}{\alpha(\Delta t)^2}[\Delta u] - \frac{1}{\alpha(\Delta t)}[^{t}\dot{u}] - \left(\frac{1}{2\alpha} - 1\right)[^{t}\ddot{u}] \tag{32}$$

$$[^{t+\Delta t}\dot{u}] = [^{t}\dot{u}] + \Delta t(1 - \beta)[^{t}\ddot{u}] + \beta\Delta t[^{t+\Delta t}\ddot{u}] \tag{33}$$

For checking of equilibrium, the following equation can be adopted,

$$[\Delta F^*]_i = [^{t+\Delta t}F_a] - ([M][^{t+\Delta t}\ddot{u}]_i + [C][^{t+\Delta t}\dot{u}]_i + [^{t+\Delta t}R]_i) \tag{34}$$

in which $[\Delta F^*]_i$ is the unbalanced residual force vector, $[^{t+\Delta t}F_a]$ is the applied nodal force, $[^{t+\Delta t}R]_i$ is the resisting force of the whole structure to be discussed in subsequent section and "i" is the number of iteration in a particular time increment. If the ratio of the Euclidean norm of $[\Delta F^*]_i$ to that of $[^{t+\Delta t}F_a]$ is less than a certain tolerance of, say, 0.1% in the present studies, equilibrium condition is assumed to have been satisfied and no further equilibrium iteration is needed. The procedure in Equation (27) for next time step is then repeated. Otherwise, the following iterative scheme is activated (see also Chui and Chan, 1993).

The residual displacement $[\Delta u]_i$ due to the unbalanced residual force $[\Delta F^*]_i$ is given by,

$$[\Delta u]_i = [K]_{eff}^{-1}[\Delta F^*]_i \tag{35}$$

The residual displacement in Equation (35) is then used to update the geometry, the displacement, the velocity and the acceleration by

$$[^{t+\Delta t}x]_{i+1} = [^{t+\Delta t}x]_i + [\Delta u]_i \tag{36}$$

$$[^{t+\Delta t}u]_{i+1} = [^{t+\Delta t}u]_i + [\Delta u]_i \tag{37}$$

$$[^{t+\Delta t}\dot{u}]_{i+1} = [^{t+\Delta t}\dot{u}]_i + \frac{\beta}{\alpha\Delta t}[\Delta u]_i \tag{38}$$

$$[^{t+\Delta t}\ddot{u}]_{i+1} = [^{t+\Delta t}\ddot{u}]_i + \frac{1}{\alpha(\Delta t)^2}[\Delta u]_i \tag{39}$$

Procedure from Equation (27) to (39) is iterated until convergence is achieved. After satisfaction of equilibrium at a particular time instance, the next time increment is marched out from Equation (27) and the process is repeated until the time history response for the desired duration is tracked.

2.5.8. Determination of Resistant Member Forces

The resisting force of a structure under deformation $[R]$ should be evaluated to check whether or not convergence is less than the desired tolerance. It is obtained by adding the incremental resisting member force due to the displacement increment $[\Delta u]$ to the last resisting force. The procedure is described as follows.

After solving the incremental nodal displacement $[\Delta u]$ about the global axis from Equation (27) or (35), the nodal displacement increments $[\Delta\psi]$ about the last local coordinate axis can be obtained directly as,

$$
\begin{aligned}
[\Delta\psi] &= [L]^T[\Delta u] \\
&= [\Delta u_1 \; \Delta v_1 \; \Delta w_1 \; \Delta\theta_1^x \; \Delta\theta_1^y \; \Delta\theta_1^z \\
&\quad \Delta u_2 \; \Delta v_2 \; \Delta w_2 \; \Delta\theta_2^x \; \Delta\theta_2^y \; \Delta\theta_2^z]^T
\end{aligned}
\tag{40}
$$

where $[L]^T$ is the transformation matrix converting the 12-displacements from the global axis to the local axis, Δu, Δv and Δw are the translational displacement increments, $\Delta\theta$ is the rotational displacement increment, the subscripts "1" and "2" are the nodes 1 and 2, the superscript "x", "y" and "z" are the local referenced axes.

Hence, the external rotation increments $\Delta_e\theta$ about the axis passing through the end nodes in the updated coordinate system can be determined by subtracting the rigid body rotation increments $\Delta\mu$ from the rotation increments $\Delta\theta$ about the last known axis. This task can be carried out by the following equations.

$$
\Delta_e\theta_1^y = \Delta\theta_1^y + \Delta\mu^y
\tag{41}
$$

$$
\Delta_e\theta_2^y = \Delta\theta_2^y + \Delta\mu^y
\tag{42}
$$

$$
\Delta_e\theta_1^z = \Delta\theta_1^z + \Delta\mu^z
\tag{43}
$$

$$
\Delta_e\theta_2^z = \Delta\theta_2^z + \Delta\mu^z
\tag{44}
$$

in which the incremental rigid body rotations are given by,

$$
\Delta\mu^y = \sin^{-1}\left(\frac{\Delta w_2 - \Delta w_1}{L}\right)
\tag{45}
$$

$$
\Delta\mu^z = \sin^{-1}\left(\frac{\Delta v_2 - \Delta v_1}{L}\right)
\tag{46}
$$

Once the incremental external rotations $\Delta_e\theta$ about the last known coordinate axes are evaluated, the corresponding internal rotations can be determined by re-arranging Equation (14) as,

$$
\begin{bmatrix} \Delta_i\theta_1^y \\ \Delta_i\theta_2^y \end{bmatrix} = \begin{bmatrix} r_1^y + \dfrac{4EI^y}{L} & \dfrac{2EI^y}{L} \\[2ex] \dfrac{2EI^y}{L} & r_2^y + \dfrac{4EI^y}{L} \end{bmatrix}^{-1} \begin{bmatrix} r_1^y & 0 \\ 0 & r_2^y \end{bmatrix} \begin{bmatrix} \Delta_e\theta_1^y \\ \Delta_e\theta_2^y \end{bmatrix} \tag{47}
$$

$$
\begin{bmatrix} \Delta_i\theta_1^z \\ \Delta_i\theta_2^z \end{bmatrix} = \begin{bmatrix} r_1^z + \dfrac{4EI^z}{L} & \dfrac{2EI^z}{L} \\[2ex] \dfrac{2EI^z}{L} & r_2^z + \dfrac{4EI^z}{L} \end{bmatrix}^{-1} \begin{bmatrix} r_1^z & 0 \\ 0 & r_2^z \end{bmatrix} \begin{bmatrix} \Delta_e\theta_1^z \\ \Delta_e\theta_2^z \end{bmatrix} \tag{48}
$$

With the external and the internal rotations at the ended connections about the local y and z axes $\Delta_e\theta$ and $\Delta_i\theta$, the corresponding relative slip rotations $\Delta\phi$ at the connections can be calculated directly as,

$$
\Delta\phi = \Delta_e\theta - \Delta_i\theta \tag{49}
$$

Therefore, the incremental moments ΔM at the connections about the last known axis can be evaluated in terms of the internal rotations $\Delta_i\theta_1$ and $\Delta_i\theta_2$ by,

$$
\Delta M_1^y = \frac{4EI^y}{L}\Delta_i\theta_1^y + \frac{2EI^y}{L}\Delta_i\theta_2^y \tag{50}
$$

$$
\Delta M_2^y = \frac{2EI^y}{L}\Delta_i\theta_1^y + \frac{4EI^y}{L}\Delta_i\theta_2^y \tag{51}
$$

The incremental moments in Equation (50) and (51) are referred to the local y-axis. Similarly, the incremental moments about the local z-axis can be obtained by replacing the superscript "y" by "z" in these equations.

The incremental torsional moment ΔM_x and the incremental axial load ΔP can be evaluated as,

$$
\Delta M_x = \frac{GJ + P \cdot \rho_0^2}{L}(\Delta\theta_2^x - \Delta\theta_1^x) \tag{52}
$$

$$
\begin{aligned}
\Delta P = \frac{EA}{L}\Bigg[&(\Delta u_2 - \Delta u_1) \\
&+ \frac{L(4\Delta_i\theta_1^y{}_i\theta_2^y - \Delta_i\theta_1^y{}_i\theta_2^y - \Delta_i\theta_2^y{}_i\theta_1^y + 4\Delta_i\theta_2^y{}_i\theta_2^y)}{30} \\
&+ \frac{L(4\Delta_i\theta_1^z{}_i\theta_2^z - \Delta_i\theta_1^z{}_i\theta_2^z - \Delta_i\theta_2^z{}_i\theta_1^z + 4\Delta_i\theta_2^z{}_i\theta_2^z)}{30} \Bigg]
\end{aligned} \tag{53}
$$

in which $(\Delta u_2 - \Delta u_1)$ and $(\Delta\theta_2^x - \Delta\theta_1^x)$ are the incremental axial deformation and torsional twist, P is the axial force in the element, A is the cross-sectional area, G is the shear modulus, J is the torsional constant, and ρ_0 is the polar radius of gyration. The second and third terms in Equation (52) are to account for the bowing effect due to the axial force.

Internal forces for each element will then be updated by the force increments in Equation (50) to (53) as,

$$[P]_{i+1} = [P]_i + [\Delta P]_i \tag{54}$$

in which $[P]$ is the internal force vector given by,

$$[P] = [\ P\ M_1^y\ M_1^z\ M_x\ M_2^y\ M_2^z\ ]^T \tag{55}$$

The resistance force vector can then be computed as,

$$[R] = \sum^{\text{Nele}} [L][T][P] \tag{56}$$

in which $[T]$ and $[L]$ are the transformation matrices relating the 6 internal forces to the 12 forces in local coordinates axes and then to the global coordinates, and "Nele" is the total number of elements. For completeness, theses matrices are listed in the Appendix for convenience of the readers. Interested readers can refer to Gere and Weaver (1965) and Meek (1971) for matrix structural analysis methods.

2.6. NUMERICAL EXAMPLES

Examples 6.1 to 6.4 are to verify the validity of the developed computer program based on the proposed method. In the remaining examples from 6.5 to 6.7, the structural response of steel frames subjected to sinusoidal, sudden, cyclic and seismic loads are investigated against the connection stiffness. Rigid connections and linear stiffness connections are also included for comparison against nonlinear connections to illustrate the importance of connection flexibility and nonlinearity. Static gravitational loads are also studied because they will reduce column stiffness via the P-Delta effects.

2.6.1. A Cantilever Beam with Suddenly Applied Uniformly Distributed Load

A cantilever beam rigidly fixed to the support is analysed in this example. The beam is subjected to a sudden uniformly disturbed load, as shown in

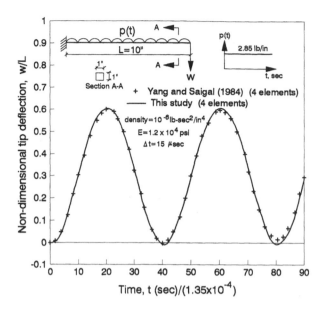

Figure 5. Large displacement dynamic response of a cantilever beam fully fixed to the support.

Figure 5. The beam is made of material with young modulus of elasticity (E) of 1.2×10^4 psi, poisson ratio (ν) of 0.2 and density (ρ) of 10^{-6} lb-sec^2/in^4. Four elements are used and a time step of 15μ second is employed. In order to simulate the fixed support of the beam, the stiffness of the spring at the support is set to $10^{10}EI/L$. The response of the tip deflection by the developed program is shown in Figure 5. The results are compared with Yang and Saigal (1984) and good agreement was obtained. It was noted that the amplitude is very large of up to 60 percent of its length, indicating the validity of the method in large deflection range.

2.6.2. A Clamped-Clamped Beam Subjected to a Sudden Load

A clamped-clamped beam under a concentrated load applied at the mid-span has been investigated by Mondkar and Powell (1977) and Yang and Saigal (1984) using the rigid joint assumption at supports. In the present study, the results for the rigid connection case are compared and the effect of the semi-rigid supports on the response is investigated. The beam is divided into

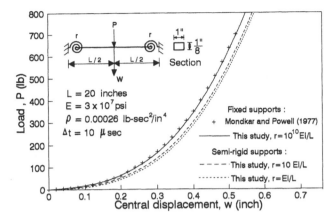

Figure 6. Nonlinear static analysis of a clamped-clamped beam with semi-rigid supports.

ten elements and its properties are shown in Figure 6. For the condition of fully rigid joints at supports, the value of the connection stiffness, r, is assigned to a very large value of $r = 10^{10} EI/L$. Since the connection between the beam and the support may not be perfectly rigid, two semi-rigid connections of stiffness of $r = 10EI/L$ and $r = EI/L$ have been assumed. The static load-deformation curves of the mid-span deflection are shown in Figure 6. For rigid connections, it was found to be very close to the results by Mondkar and Powell (1977) and Yang and Saigal (1984). It was noted that the beam stiffness (i.e. the slope of the curves) increases with the load due to the presence of the considerable axial tension in the beam in the large deflection range.

The dynamic behaviour of the beam with various connection stiffness is further considered. A sudden load of 640 lb is applied at the mid-span. The Newmark method with $\alpha = 0.25$ and $\beta = 0.5$ was employed and a time step of 10μ seconds was chosen. The results are shown in Figure 7 and found to be in close agreement with those by the other researchers. The initial responses for the semi-rigid joints are similar to the fully clamped case but subsequent paths are deviated from the rigid joint case. As expected, the beam with softer connections has a longer period and a larger peak deflection.

2.6.3. Large Deflection Dynamic Analysis of a 45° Curved Cantilever Beam

The static analysis of a cantilever 45° bend subjected to a concentrated end load has been studied by Bathe and Bolourchi (1979), Cardona and

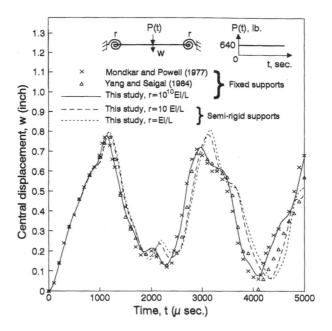

Figure 7. Dynamic response of the clamped-clamped beam with semi-rigid supports.

Geradin (1988) and Kouhia and Tuomala (1993). This example is a three-dimensional large deflection problem. In order to validate the proposed theory, this problem is re-analysed and semi-rigid connections at the support are also studied for comparison. The layout of the cantilever is shown in Figure 8. Two rotational connection springs about the local axes, $y - y$ and $z - z$, are assumed at the support end. The semi-rigid connection of stiffness is taken as $r = 10EI/L$ and the bend is modelled by eight straight elements. The static results are plotted in Figure 8. They agree well with the results by Kouhia and Tuomala (1993). The tip positions at various load levels tabulated in Table 1 are also close to the researchers.

Since its dynamic behaviour has not been studied yet, the problem is further extended to dynamic analysis. Material density is assumed to be 2.54×10^{-4} lb-sec^2/in^4. A concentrated load of 300 lb suddenly applied at the tip is studied and the time history of the tip is plotted in Figure 9. From the results, the bend with semi-rigid connection has a larger deflection amplitude and a longer response period when compared to those for rigid connection case. It can been seen that semi-rigid connection can significantly affect the response

Table 1. Comparison of static results for the cantilever 45 degree bend

Load level (lb)	This study	Tip position (inch) Kouhia and Tuomala (1993)	Cardona and Geradin (1988)
0	(70.71, 0., 29.29)	(70.71, 0., 29.29)	(70.71, 0., 29.29)
300	(58.77, 40.17, 22.29)		(58.64, 40.35, 22.14)
450	(52.22, 48.48, 18.56)		(52.11, 48.59, 18.38)
600	(47.12, 53.45, 15.75)	(47.01, 53.50, 15.69)	(47.04, 53.50, 15.55)

especially in dynamic analysis. To this end, the present developed computer program is demonstrated to have capable of handling the transient response of structures undergoing large deflection in three-dimensional space.

2.6.4. Experimental Verification by a Single-Bay Two-Storey Frame

This example is to verify the present proposed connection modelling algorithm with experimental test results. A total of ten flexibly connected steel frames were tested by Stelmack *et al.* (1986). A layout of these frames is shown in Figure 10. 1/2 inch thick angle connections were used in the test. 14 monotonic moment-rotation test results of such connections were reported in their references, and the upper and lower bounds of the scattered data were shown in Figure 11. The trilinear model was used by Stelmack *et al.* (1986) whilst the Richard-Abbott model is used in the present study. Youssef-Agha *et al.* (1989) also studied the seismic response of a similar frame. In their paper, a bilinear model, particularly for the top and seat angle connection type, was proposed and expressed in terms of the connection details such as the dimensions, geometries and arrangement of the connection components. However, this model is not general for other connection types. In the test by Stelmack *et al.* (1986), the frame was subjected to cyclic lateral loads which began with load applied to the first storey of ±1 kip and increased or decreased by 1 kip increment up to ±5 kips while the second-storey load is always one-half of the first storey load. The load history is plotted on Figure 12.

The experimental and analytical results of the load-deflection curves and the connection moment-rotation loops are compared and plotted separately for each cycle in Figures 13 and 14. It can be seen that the predicted results by the Richard-Abbott model is close to the experimental results and is smoother than the results by the trilinear model. This shows the proposed theory can predict an accurate and smooth deflection curves for the frame.

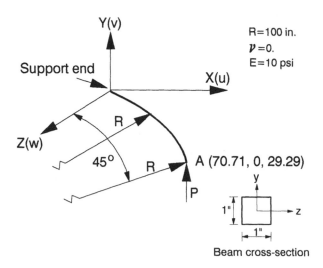

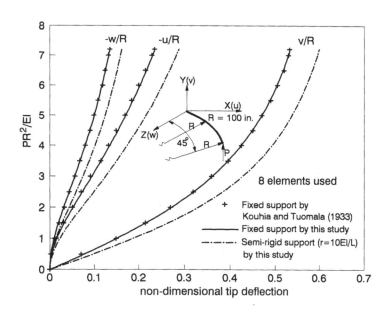

Figure 8. Three-dimensional large deflection of a 45° circular bend.

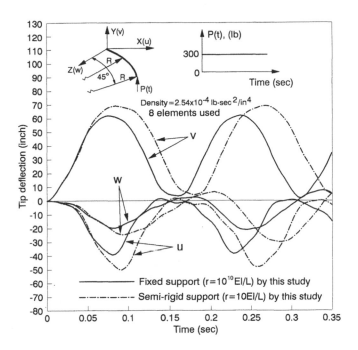

Figure 9. Dynamic behaviour of a 45° circular bend.

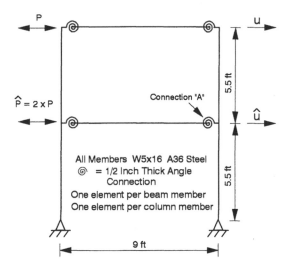

Figure 10. A one-bay two-storey frame.

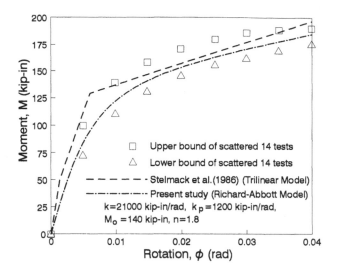

Figure 11. Moment-rotation curves of 1/2 inch thick angle connection.

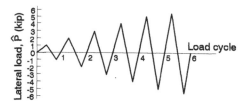

Figure 12. Lateral load history.

The next study is to investigate the influence of the choice of connection models. The frame with rigid, linear and nonlinear connection types are assumed. For the nonlinear connections, three models, namely the Richard-Abbott model, the Ramberg-Osgood model and the exponential model, are used. The representation of these models on $1/2''$ angle connection is shown in Figure 15. The frame is assumed to be subjected to a guest wind loads which push and pull the frame in one second and then disappear. The time histories of the deflections at the second storey of the frames are depicted in Figure 16. It can be seen that the frame with semi-rigid connections has a larger amplitude and a longer period when compared with the rigid connection case. Also, the frame with nonlinear connections dampens and exhibits a non-recoverable

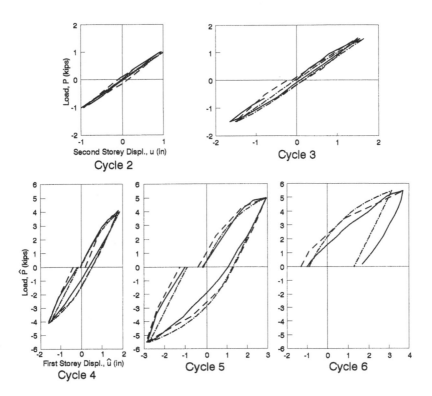

Figure 13. Load-deflection curves.

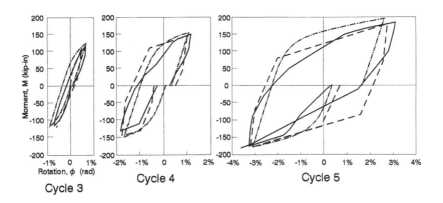

Figure 14. Moment-rotation loops at connection "A".

——————————— Tested results by Stelmack et al. (1986)
— — — — — — — Theory predicted by Stelmack et al. (1986)
—·—·—·—·—·— Predicted results by present theory

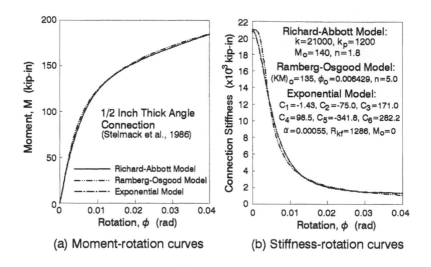

Figure 15. Models representing 1/2 inch thick connection.

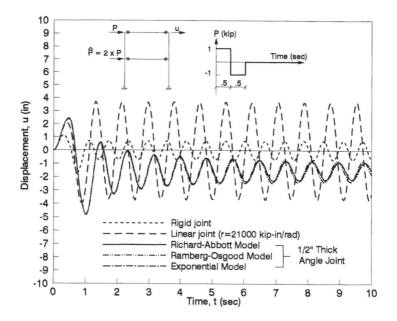

Figure 16. Transient response of frames with different joints.

deflection due to the presence of permanent rotations at connections. This feature was not observed in a linear joint model. In contrary to the nonlinear connection case, the frame with either rigid or linear connections will not dampen. It was also noted that the response of the frame simulated by the three nonlinear connection models are very close to each other, indicating that the choice of these three connection models is not important when the connection behaviour is represented appropriately by them. Moreover, it also shows that the proposed algorithm for simulating hysteretic and cyclic moment-rotation loops at connections can describe for different connection models. As shown in the previous study, the coarse model based on the simple trilinear function produces a curve with kinks. This slope discontinuity may not be desired in a computer analysis and thus the linearized model is not recommended.

2.6.5. A Simple Portal Frame Subjected to Sinusoidal Loads

Figure 17 shows an unbraced portal frame carrying two nodal masses each of 5000 kg. The dynamic response of the frame with different connection types and under various sinusoidally exciting loads is investigated. Flush end-plate connection which was tested by Ostrander (1970) and then modelled by an exponential function by Lui and Chen (1988), as shown in Figure 18 is used. Linear and rigid joint types are also considered for comparison. Based on the modal analysis method by Chan (1994), the fundamental natural frequencies of the portal frame are computed as 7.72 rad/sec and 9.20 rad/sec for the linear and rigid joint cases respectively. Due to the variation of connection stiffness under loads, the portal frame with nonlinear flush end-plate joints shows a frequency lower than 7.72 rad/sec. The portal frame is assumed to be subjected to a lateral force of magnitude of 30kN with five exciting frequencies (5.5, 7.72, 8.45, 9.20 and 11.52 rad/sec) and an impulse load of the same magnitude for 0.3 second.

The results of analysis are shown in Figure 19. As expected, the frame with linear and rigid joints will resonate at excitation of 7.72 and 9.20 rad/sec respectively, as shown in Figures 19(b) and (d). However, the frame with nonlinear connections does not resonate but it stabilizes by damping due to the energy dissipation in the hysteretic moment-rotation loops at connections, as shown in Figure 20. It was observed that the amplitude of deflection of the semi-rigid frame is greater with a lower excitation force when compared with the rigid joint case, as shown in Figure 19(a). It is because the exciting frequency of the applied force is closer to the natural frequency of the semi-rigid frame. Thus, the deflection under this dynamic load is magnified. Similarly, if the exciting frequency is high, the deflection of the rigid frame shown in Figure 19(e) is greater than the other two semi-rigid

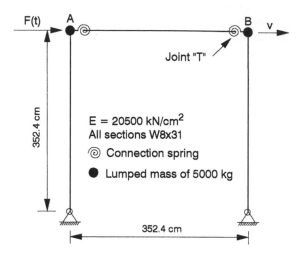

Figure 17. Layout of a portal frame.

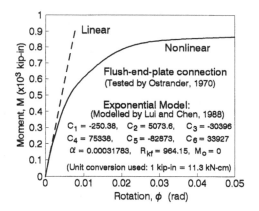

Figure 18. Monotonic moment-rotation curve of flush end-plate connection.

frames. In Figure 19(c), when the load frequency lies between the natural frequencies of semi-rigid and rigid frames, their deflections will be similar. From the results of analysis, the semi-rigid frame with nonlinear connections will magnify displacements at low excitation force but dampened at high excitation, depending on proximity between the vibrating frequency and the structural natural frequency. Under a short time impact load in Figure 19(f),

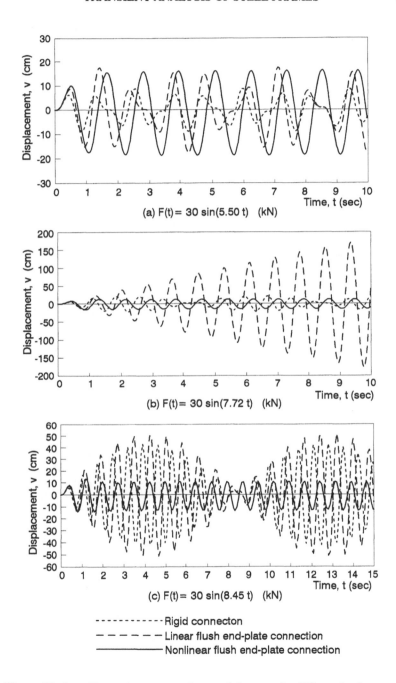

Figure 19a,b,c. Dynamic response of a portal frame under different loads.

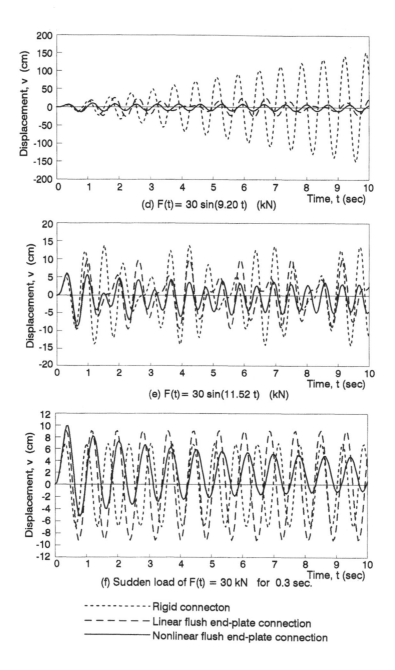

(d) F(t) = 30 sin(9.20 t)   (kN)

(e) F(t) = 30 sin(11.52 t)   (kN)

(f) Sudden load of F(t) = 30 kN   for 0.3 sec.

----------- Rigid connecton
− − − − − Linear flush end-plate connection
———— Nonlinear flush end-plate connection

Figure 19d,e,f. Dynamic response of a portal frame under different loads.

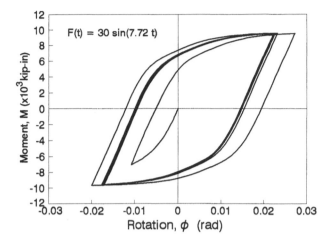

Figure 20. Stable and reproducible loops of hysteretic $M - \phi$ relationship at joint "T" for case (b).

it can be seen that damping appears in the frame with nonlinear connections but does not occur in the frame with neither rigid nor linear joints.

2.6.6. Gravity Load Effects on a Single-Bay Two-Storey Frame

To predict a more realistic behaviour of a structure, gravity loads (e.g. floor slabs and furniture) statically acting on the structure should be included in the dynamic analysis. It is because gravity loads can induce axial forces on columns and thus reduce the column stiffness. Also, an overturning moment due to the P-Delta effect will be produced by the gravity loads. In order to study these effects on structural response, gravity loads are taken into account. These effects have been studied by Chui and Chan (1995).

An unbraced two-storey frame with static gravity loads, 36.6 and 73.2kN, are studied and its geometry is shown in Figure 21(a). The loads are assumed to be induced by a mass of 14920 kg on each floor, as shown in Figure 21(b), and are considered as lumped masses in the dynamic analysis. Linear and nonlinear flush end-plate and rigid connection types are assumed. The flush end-plate connection modelled by the exponential model by Lui and Chen (1988) is assumed and plotted in Figure 18. Two elements per beam member and one element per column are used for modelling, and a time step of $\Delta t = 0.05$ second is chosen. The static load-deformation response of the

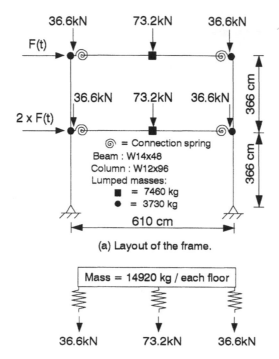

(a) Layout of the frame.

(b) Equivalent gravity loads induced by a heavy mass on each floor.

Figure 21. A two-storey height portal frame.

frames is studied and plotted in Figure 22. The load carrying capacity is considerably affected by the nonlinear behaviour of connections. To test the reproducibility of hysteretic moment-rotation loops at connections, the frame is subjected to cyclic loads, and the response of displacement and connection $M - \phi$ loops are traced in Figure 23. The deflection initially oscillates between ±7cm at transient state but later oscillates steadily between ±5cm. It can be seen in Figure 23(b) that the hysteretic loops are highly reproducible and stable. During this stable response, the amount of the input energy by the exciting loads and the energy dissipation at connections are found to be the same and approximately equal to 320 Joules per each cycle. In Figure 24, the same frame is studied against the presence and the absence of gravity loads and under the sudden loads. Various connection types are also considered. It was found that the deflection amplitudes of the frames accounting for gravity loads is larger than those without gravity loads.

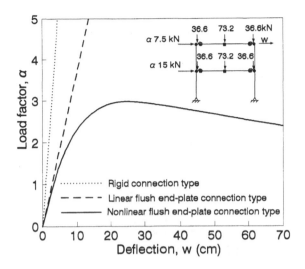

Figure 22. Static analysis of a portal frame with different connection types.

It is due to the column softening and the P-Delta effects. Because of the formation of permanent rotations at the connections, there exists a shift of deflection response in the flexibly jointed frames. As indicated in the previous example, nonlinear connections increase the structural response frequency whilst hysteretic damping induced at the connections decreases the successive amplitudes of the response due to dissipation of energy.

2.6.7. Seismic Response of a Four-Bay Five-Storey Frame

In a seismicity zone, it is necessary to insure buildings possess adequate strength to prevent total collapse under severe earthquakes or serious damage under moderate earthquakes. Since flexible connections can greatly alter the behaviour of structures, it is interesting to see their significance on seismic response.

The layout of a realistic four-bay five-storey steel frame located in Shanghai, China, is examined and a similar model with geometry and property is shown in Figure 25. Wong (1995) studied the static load-deformation analysis of the structure with flush end-plate connection type which was assumed to be the same type represented by the exponential model by Lui and Chen (1988). In present study, the flush end-plate and rigid connection types are considered for comparison. Five elements per

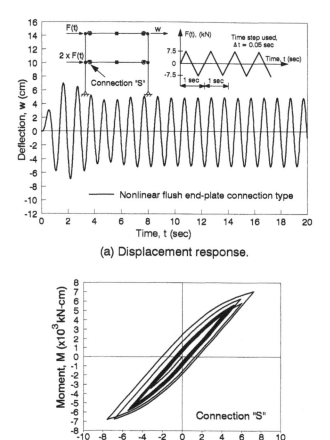

(a) Displacement response.

(b) Response of the hysteretic M-ϕ loops at connection "S".

Figure 23. Dynamic behaviour of a frame with nonlinear flush end-plate connections under cylic loads.

beam member and one element per column are employed in modelling of the structure. Gravity loads are also included and they are considered as additional lumped masses at the nodes on the beam. The frame is assumed to be subjected to the first ten seconds of the 1940 EI Centro N-S earthquake component motion recorded in USA. The records of the ground motion can be found from some standard texts, such as Dowrick (1988) and Paz (1991). The time-history of the ground motion was measured as shown in Figure 26

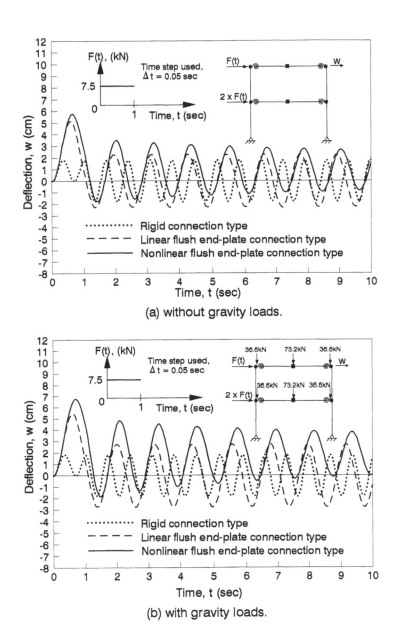

Figure 24. Dynamic response of steel frames with various connection types.

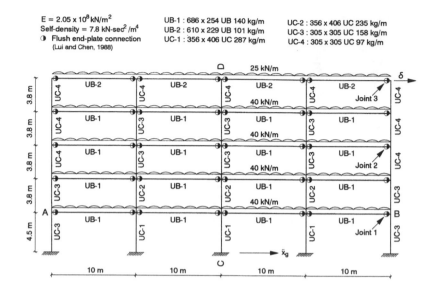

Figure 25. Geometry and property of a four-bay five-story steel frame.

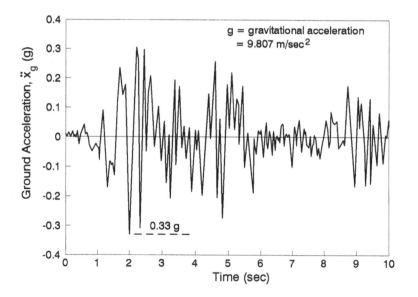

Figure 26. Accelerogram for El Centro (1940) NS component.

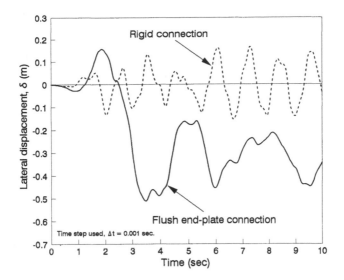

Figure 27. Time history of displacement response.

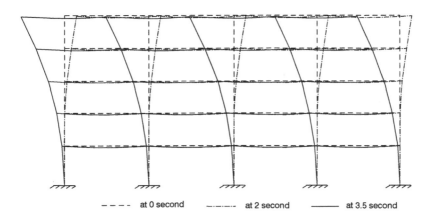

Figure 28. Ten times enlarged deformed shapes of the frame with the nonlinear flush end-plate connection type at the specified time.

and the peak ground acceleration was 0.33g at about 2 seconds. The transient response of the lateral displacement, δ, at the top of the right-most column is predicted and traced in Figure 27. The deflected shapes of the structure with the nonlinear connections at 2 and 3.5 seconds are also depicted in Figure 28.

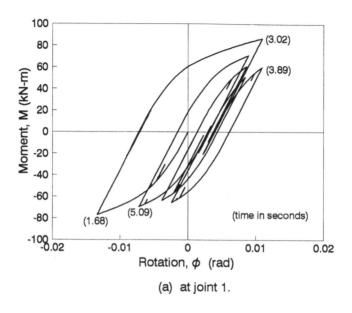

(a) at joint 1.

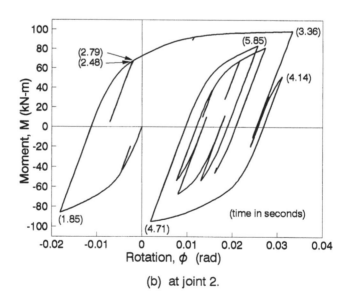

(b) at joint 2.

Figure 29a,b. Response of hysteretic $M - \phi$ loops of flush end-plate connections up to ten seconds.

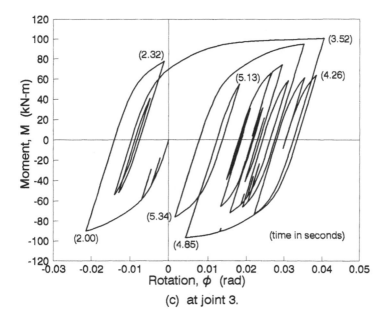

Figure 29c. Response of hysteretic $M - \phi$ loops of flush end-plate connections up to ten seconds.

It is recalled that the ground acceleration is in absolute (i.e. relative to a stationary point) while the deflection of the frame is relative to the ground. Compared with the rigid jointed frame, the flexibly connected frame deforms suddenly to a peak value of about –0.5m at near 3.5 second and then oscillates with this permanent deflection. It indicates that connections possess large deformed rotation after the peak ground acceleration. The hysteretic $M - \phi$ loops at joint 1 to 3 are traced in Figure 29 and marked with time records at turning points for clarity. It can be seen that the connections undergo severe rotational deformations during the period from 2 to 3.5 seconds. The lateral displacement envelope of the frame at different floor levels are plotted in Figure 30. The maximum deflection for the flush end-plate joint case is approximately three times that for the rigid joint case. It is interesting to note that the backward deflection is much larger than the forward deflection due to permanent rotations formed at the connections of the flexibly connected frame. The bending moment envelopes for the beam A-B and the column C-D are also plotted in Figure 31. Due to the moment being limited by the capacity of the semi-rigid connections, the bending moments at the ends of beam members (i.e. adjacent to columns) are limited to about only one tenth of that

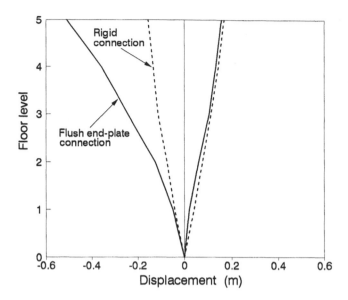

Figure 30. Lateral displacement envelope of the frame up to ten seconds.

of the rigid jointed frame. At the mid-span of beam, the bending moment for the semi-rigid jointed frame is slightly higher because less bending moments are transferred and distributed to the columns through the flexible joints. For the column C-D, it can be seen that less bending moments are induced in the flexibly jointed frame. Based on this example, it can be deduced that flexible connections increase the storey drift but reduce the base shear force significantly under seismic loads. This represents a potentially important role played by connections during earthquake and also on the design of the structure and its foundation.

2.7. CONCLUSIONS

A robust, stable and reliable incremental-iterative displacement-based finite element method for nonlinear large deflection dynamic analysis of steel structures accounting for geometric and connection nonlinearities is presented in this chapter. The proposed method was tested and found to be accurate and valid by comparison with a number of well-known benchmark problems and experimental results. The dynamic response of a steel structure is known to be significantly affected by nonlinear behaviour of beam-column

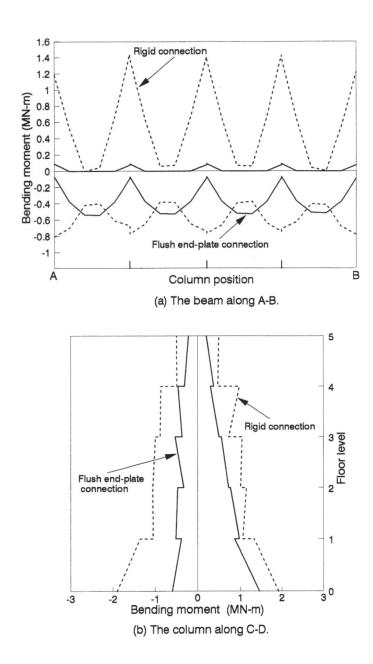

(a) The beam along A-B.

(b) The column along C-D.

Figure 31. Bending moment envelopes up to ten seconds.

connections. Apart from the viscous and inelastic damping, the generating process of hysteretic loops of moment-rotation relationship at connections is demonstrated to introduce a major damping to flexibly jointed structures. The connection hysteretic damping will increase the structural energy absorption capacity which is particularly important for a structure to survive under severe seismic loads.

The dynamic response against the choice of connection models is shown to be insensitive if the nonlinear behaviour of a connection is fitted properly. This means that the time-consuming procedure for converting model parameters from one model to another is not necessary if the moment-rotation relationship of a connection is well represented. Transient responses for the Ramberg-Osgood, the Richard-Abbott and the exponential models were compared and found to give very close results.

The P-Delta effect due to gravity loads was noted to be significant and should be included in the dynamic analysis. Furthermore, based on the numerical examples, it can be deduced that flexible connections will increase the storey drift but reduce the base shear forces under seismic loads. To this end, a flexible connection should be considered as an important structural element affecting the dynamic response considerably and must be included in the dynamic analysis.

ACKNOWLEDGEMENTS

The authors wish to thank Professor M. Anson, Professor J.M. Ko and Dr. A.K.W. So for their valuable discussions and comments on this chapter. The effort of Miss Freda Chow in preparing this manuscript is greatly appreciated. Finally, the financial support by the Research Grant Council of the Hong Kong Government and the Hong Kong Polytechnic University is gratefully acknowledged (RGC Project No. 340/938).

References

1. Al-Bermani, F.G.A. and Kitipornchai, S., 1992, "Elasto-Plastic Nonlinear Analysis of Flexibly Jointed Space Frames", *J. Struct. Div.*, ASCE, **118**(1), 108–127.
2. American Institute of Steel Construction, 1986, *LRFD Load and Resistance Factor Design Specification for Structural Steel Buildings*, AISC, Chicago.
3. Anderson, D. and Benterkia, Z., 1991, "Analysis of Semi-Rigid Frames and Criteria for their Design", *J. Construct. Steel Research*, **18**, 227–237.
4. Ang, K.M. and Morris, G.A., 1984, "Analysis of Three-Dimensional Frames with Flexible Beam-Column Connections", *Can. J. Civil Engng.*, **11**, 245–254.
5. Arbabi, F., 1982, "Drift of Flexibly Connected Frames", *Comp. and Struct.*, **15**(2), 103–108.
6. Azizinamini, A. and Radziminski, J.B., 1989, "Static and Cyclic Performance of Semirigid Steel Beam-to-Column Connections", *J. Struct. Div.*, ASCE, **115**(12), 2979–2999.

7. Bathe, K.J., 1996, *Finite Element Procedures*, Prentice Hall, New Jersey.
8. Bathe, K.J. and Bolourchi, S., 1979, "Large Displacement Analysis of Three-Dimensional Beam Structures", *Int. J. Num. Methods in Engng.*, **14**, 961–986.
9. Bell, W.G., Chesson, E.J. and Munse, W.H., 1958, *Static Tests of Standard Riveted and Bolted Beamd-to-Column Connections*, Univ. of Illinois Engng. Experiment Station, Urban, Ill.
10. British Standard Institution, 1985, *BS5950: Part I: Structural Use of Steelwork in Building*, BSI, London, England.
11. Cardona A. and Geradin M., 1988, "A Beam Finite Element Non-linear Theory with Finite Rotations", *Int. J. Num. Methods in Engng.*, **26**, 2403–2438.
12. Carpenter, L.D. and Lu, L.W., 1973, "Reversed and Repeated Load Tests of Full-Scale Steel Frames", Bulletin No. **24**, *American Iron and Steel Institute*, New York, April.
13. Chan, S.L., 1994, "Vibration and Modal Analysis of Steel Frames with Semi-Rigid Connections", *Engng. Struct.*, **16**(1), 25–31.
14. Chan, S.L. and Chui, P.P.T., 1993, "Nonlinear Dynamic Analysis of Roof Trusses Connected by Rigid, Pinned and Semi-Rigid Joints", *Proceedings of Fourth Int. Conference on Space Struct.*, U.K., (1), pp. 406–412.
15. Chan, S.L. and Ho, G.W.M., 1994, "Nonlinear Vibration Analysis of Steel Frames with Semi-Rigid Connections", *J. Struct. Div.*, ASCE, **120**(4), 1075–1087.
16. Chan, S.L. and Yau, C.Y., 1992, "Nonlinear Analysis of Roof Truss Composed of Pinned, Rigid and Semi-rigid Jointed Members", *Innovative Large Span Structures, IASS-CSCE International Congress 1992*, Vol. **2**, 175–185.
17. Chen, W.F. and Chan, S.L., 1994, "Second Order Inelastic Analysis of Steel Frames by Personal Computers", *J. of Struct. Engng.*, **21**(2), 99–106, July.
18. Chen, W.F. and Kishi, N., 1989, "Semirigid Steel Beam-to-Column Connections: Data Base and Modeling", *J. Struct. Div.*, ASCE, **115**(1), 105–119.
19. Chen, W.F. and Zhou, S.P., 1987, "Inelastic Analysis of Steel Braced Frames with Flexible Joints", *Int. J. Solids Structures*, **23**(5), 631–649.
20. Chui, P.P.T. and Chan, S.L., 1995, "Transient Response of Semi-Rigid Steel Frames", *Proceedings of Int. Conference on Struct. Dynamics, Vibration, Noise and Control*, Hong Kong, (1), 191–196.
21. Clough, R.W. and Penzien, J., 1975, *Dynamics of Structures*, McGraw-Hill, New York.
22. Clough, R.W. and Tang, D., 1975, *Earthquake Simulator Story of Steel Frame Structure, Vol. I: Experimental Results*, Report No. UCB/EERC-75/06, Earthquake Engng. Research Center, Univ. of California, Berkeley.
23. Cosenza, E., Luca, D.A. and Faella, C., 1989, "Elastic Buckling of Semi-Rigid Sway Frames", Chapter 8 in *Structural Connections, Stability and Strength*, edited by R. Narayanan, Elsevier Applied Science, pp. 253–295.
24. Cox, M.G., 1972, "The Numerical Evaluation of B-Splines", *J. Inst. Math. Applications*, **10**, 134–139.
25. Davison, J.B., Kirby, P.A. and Nethercot, D.A., 1987, "Rotational Stiffness Characteristics of Steel Beam-to-Column Connections", *J. Construct. Steel Research*, **8**, 17–54.
26. Dowrick, D.J., 1988, *Earthquake Resistant Design*, 2nd Ed., John Wiley and Sons, Singapore.
27. Eurocode 3, 1988): *Common Unified Rules for Steel Structures*, Commission of the European Communities, Brussels, Redraft.
28. Frye, M.J. and Morris, G.A., 1975, "Analysis of Flexibly Connected Steel Frames", *Can. J. Civil Engng.*, **2**(3), 280–291.
29. Gao, L. and Haldar, A., 1995, "Nonlinear Seismic Analysis of Space Structures with Partially Restrained Connections", *Microcomputers in Civil Engng.*, **10**, 27–37.
30. Gere, J.M. and Weaver, W.J., 1965, *Analysis of Framed Structures*, Van Nostrand Reinhold, New York.
31. Gerstle, K.H., 1988, "Effect of Connections on Frames", *J. Construct. Steel Research*, **10**, 241–267.

32. Goto, Y. and Chen, W.F., 1987, "On the Computer-Based Design Analysis for the Flexibly Jointed Frames", *J. Construct. Steel Research*, **8**, 203–231.

33. Goverdhan, A.V, 1984, *A Collection of Experimental Moment-Rotation Curves and Evaluation of Prediction Equations for Semi-Rigid Connections*, Thesis for Master of Science, Vanderbilt Univ., Nashville, Tenn.

34. Ho, W.M.G. and Chan, S.L., 1991, "Semibifurcation and Bifurcation Analysis of Flexibly Connected Steel Frames", *J. Structural Engineering*, ASCE, **117**(8), 2299–2319.

35. Ho, W.M.G. and Chan, S.L., 1993, "An Accurate and Efficient Method for Large Deflection Inelastic Analysis of Frames with Semi-Rigid Connections", *J. Construct. Steel Research*, **26**, 171–191.

36. Jaspart, J.P., 1988, "Extending of the Merchant-Rankine Formula for the Assessment of the Ultimate Load of Frames with Semi-rigid Joints", *J. Construct. Steel Research*, **11**, 283–312.

37. Johnstone, N.D. and Walpole, W.R., 1981, *Bolted End Plate Beam to Column Connections under Earthquake Type Loading*, Research Report 81–7, Dept. of Civil Engng., Univ. of Canterbury, Christchurch, New Zealand, Sept.

38. Jones, S.W., Kirby, P.A. and Nethercot, D.A., 1980, "Effect of Semi-Rigid Connections on Steel Column Strength", *J. Construct. Steel Research*, **1**, 38–46.

39. Kawashima, S. and Fujimoto, T., 1984, "Vibration Analysis of Frames with Semi-Rigid Connections", *Comp. and Struct.*, **19**, 85–92.

40. Kishi, N. and Chen, W.F., 1986a, *Data Base of Steel Beam-to-Column Connections*, CE-STR-86–26, School of Civil Engng., Purdue Univ., W. Lafayette, Ind.

41. Kishi, N. and Chen, W.F., 1986b, *Steel Connection Data Bank Program*, CE-STR-86–18, School of Civil Engng., Purdue Univ., W. Lafayette, Ind.

42. Korol, R.M., Ghobarah, A. and Osman, A., 1990, "Extended End-Plate Connections Under Cyclic Loading: Behaviour and Design", *J. Construct. Steel Research*, **16**, 253–280.

43. Kouhia, R. and Tuomala M., 1993, "Static and Dynamic Analysis of Space Frames using Simple Timoshenko Type Elements", *Int. J. Num. Methods in Engng.*, **36**, 1189–1221.

44. Krawinkler, H., Bertero, V.V. and Popov, E.P., 1971, *Inelastic Behaviour of Steel Beam-to-Column Subassemblages*, Report No. UBC/EERC-71/07, Earthquake Engng. Research Center, Univ. of California, Berkeley.

45. Leger, P. and Dussault, S., 1992, "Seismic-Energy Dissipation in MDOF Structures", *J. Struct. Div.*, ASCE, **118**(5), 1251–1269.

46. Lui, E.M. and Chen, W.F., 1987, "Steel Frame Analysis with Flexible Joints", *J. Construct. Steel Research*, **8**, 161–202.

47. Lui, E.M. and Chen, W.F., 1988, "Behavior of Braced and Unbraced Semi-Rigid Frames", *Int. J. Solids Struct.*, **24**(9), 893–913.

48. Meek, J.L., 1971, *Matrix Structural Analysis*, McGraw-Hill, N.W.

49. Mondkar, D.P. and Powell, G.H., 1977, "Finite Element Analysis of Non-Linear Static and Dynamic Response", *Int. J. Num. Methods in Engng.*, **11**, 499–520.

50. Moore, D.B., Nethercot, D.A. and Kirby, P.A., 1993a, "Testing Steel Frames at Full Scale", *Struct. Engineer*, **71**, 418–427.

51. Moore, D.B., Nethercot, D.A. and Kirby, P.A., 1993b, "Testing Steel Frames at Full Scale: Appraisal of Results and Implications for Design", *Struct. Engineer*, **71**, 428–435.

52. Nader, M.N. and Astaneh, A., 1991, "Dynamic Behaviour of Flexible, Semirigid and Rigid Steel Frames", *J. of Construct. Steel Research*, **18**, 179–192.

53. Nethercot, D.A., 1985, *Steel Beam-to-Column Connections - A Review of Test Data*, Construction Industry Research and Information Association, London, England.

54. Newmark, N.M., 1959, "A Method of Computation for Structural Dynamics", *J. Engng. Mech. Div.*, ASCE, **85**(3), 67–94.

55. Ostrander, J.R., 1970, *An Experimental Investigation of End-Plate Connections*, Master's Thesis, Univ. of Saskatchewan, Saskatoon, Saskatchewan, Canada.

56. Paz, M., 1991, *Structural Dynamics: Theory and Computation*, 3rd Ed., Van Nostrand Reinhold, New York.

57. Popov, E.P., 1983, *Seismic Moment Connections for Moment-Resisting Steel Frames*, Report No. UCB/EERC-83/02, Earthquake Engng. Research Center, Univ. of California, Berkeley, California, January.

58. Popov, E.P. and Bertero, V.V., 1973, "Cyclic Loading of Steel Beams and Connections", *J. Struct. Div.*, ASCE, **99**(6), 1189–1204.

59. Popov, E.P. and Pinkney, R.B., 1969, "Cyclic Yield Reversal in Steel Building Connections", *J. Struct. Div.*, ASCE, **95**(3), 327–353.

60. Popov, E.P. and Stephen, R.M., 1970, *Cyclic Loading of Full-Size Steel Connections*, Report No. UCB/EERC-70/03, Earthquake Engng. Research Centre, Univ. of California, Berkeley, July. (Republished as Bulletin No. 21, American Iron and Steel Institute, 1972.)

61. Ramberg, W. and Osgood, W.R., 1943, *Description of Stress-Strain Curves by Three Parameters*, Tech. Report No. 902, Nat. Advisory Committee for Aeronautics, Washington, D.C.

62. Rathbun, J.C., 1936, "Elastic Properties of Riveted Connections", *ASCE Trans.*, **101**(1933), 524–563.

63. Richard, R.M., and Abbott, B.J., 1975, "Versatile Elastic-Plastic Stress-Strain Formula", *J. Engng. Mech. Div.*, ASCE, **101**(4), 511–515.

64. Savard, M., Beaulieu, D. and Fafard, M., 1994, "Nonlinear Finite Element Analysis of Three-Dimensional Frames", *Can. J. Civil Engng.*, **21**, 461–470.

65. Shi, G. and Atluri, S.N., 1989, "Static and Dynamic Analysis of Space Frames with Nonlinear Flexible Connections", *Int. J. Num. Methods in Engng.*, **28**, 2635–2650.

66. Sivakumaran, K.S., 1988, "Seismic Response of Multi-Storey Steel Buildings with Flexible Connections", *Engng. Struct.*, **10**, 239–248.

67. Sommer, W.H., 1969, *Behaviour of Welded Header Plate Connections*, Master's Thesis, Univ. of Toronto, Ont., Canada.

68. Stelmack, T.W., Marley, M.J. and Gerstle, K.H., 1986, "Analysis and Tests of Flexibly Connected Steel Frames", *J. Struct. Engng.*, ASCE, **112**(7), 1573–1588.

69. Tsai, K.C. and Popov, E.P., 1990, "Cyclic Behavior of End-Plate Moment Connections", *J. Struct. Div.*, ASCE, **116**(11, 2917–2930.

70. Vedero, V.T. and Popov, E.P., 1972, "Beam-Column Subassemblages under Repeated Loading", *J. Struct. Div.*, ASCE, **98**(5), 1137–1159.

71. Wilson, W.M. and Moore, H.F., 1917, *Tests to Determine the Rigidity of Riveted Joints in Steel Structures*, Bulletin No. 104, Engng. Experiment Station, Univ. of Illinois, Urbana.

72. Wong, H.P., 1995, *A Study on the Behaviour of Realistic Steel Buildings using Design and Rigorous Approaches*, Thesis submitted for partial fulfillment for the Degree of Master of Science, Hong Kong Polytechnic Univ., Hong Kong.

73. Yang, T.Y. and Saigal, S., 1984, "A Simple Element for Static and Dynamic Response of Beams with Material and Geometric Nonlinearities", *Int. J. Num. Methods in Engng.*, **20**, 851–867.

74. Yau, C.Y. and Chan, S.L., 1994, "Inelastic and Stability Analysis of Flexibly Connected Steel Frames by Springs-in-Series Model", *J. Struct. Engng.*, ASCE, **120**(10), 2803–2819.

75. Young, C.R. and Jackson, K.B., 1934, "The Relative Rigidity of Welded and Riveted Connections", *Can. J. of Research*, **11**(1–2), 62–134.

76. Youssef-Agha, W., Aktan, H.M. and Olowokere, O.D., 1989, "Seismic Response of Low-Rise Steel Frames", *J. of Struct. Div.*, ASCE, **115**(3), 594–607.

APPENDIX 1

The 12×6 Transformation Matrix, $[T]$, which is used to transform a 6×1 vector to a 12×1 vector:

$$[T] = \begin{bmatrix} -1 & 0 & 0 & 0 & 0 & 0 \\ 0 & 0 & 1/L & 0 & 0 & 0 \\ 0 & -1/L & 0 & 0 & -1/L & 0 \\ 0 & 0 & 0 & -1 & 0 & 0 \\ 0 & 1 & 0 & 0 & 0 & 0 \\ 0 & 0 & 1 & 0 & 0 & 0 \\ 1 & 0 & 0 & 0 & 0 & 0 \\ 0 & 0 & -1/L & 0 & 0 & -1/L \\ 0 & 1/L & 0 & 0 & 1/L & 0 \\ 0 & 0 & 0 & 1 & 0 & 0 \\ 0 & 0 & 0 & 0 & 1 & 0 \\ 0 & 0 & 0 & 0 & 0 & 1 \end{bmatrix}_{12 \times 6}$$

The 12×12 Direction Cosine Matrix, $[L]$, which transforms a 12×1 vector in the local coordinate axes to the global coordinate axes:

$$[L] = \begin{bmatrix} [D] & 0 & 0 & 0 \\ 0 & [D] & 0 & 0 \\ 0 & 0 & [D] & 0 \\ 0 & 0 & 0 & [D] \end{bmatrix}_{4 \times 4} \qquad \text{(partitioned)}$$

in which

$$[D] = \begin{bmatrix} C_x & \frac{-(C_x+C_y)}{Q} & \frac{-C_z}{Q} \\ C_y & Q & 0 \\ C_z & \frac{-(C_y+C_z)}{Q} & \frac{C_x}{Q} \end{bmatrix}_{3 \times 3}$$

$C_x = (X_2 - X_1)/L$; $C_y = (Y_2 - Y_1)/L$; $C_z = (Z_2 - Z_1)/L$; $Q = (C_x^2 + C_y)^{1/2}$, L is the length of the element; X, Y and Z are the coordinates; and the subscript "1" and "2" are the reference to the node 1 and 2 of the element respectively.

3 SOME TECHNIQUES IN OPTIMAL CONTROL OF BUILDING FRAMES

CHIH-CHEN CHANG[1], HENRY T.Y. YANG[2] and SHAN-MIN SWEI[1]

[1]*Department of Mechanical Engineering, Hong Kong University of Science and Technology, Clear Water Bay, Kowloon, Hong Kong*
[2]*Department of Mechanical and Environmental Engineering and Chancellor, University of California, Santa Barbara, CA 93106*

3.1. INTRODUCTION

In the wake of the devastating destruction caused by some of the recent natural hazards, such as, the Mexico Earthquake in 1985, the Loma Prieta Earthquake in 1989, the Andrew and Iniki Hurricanes which hit Florida and Hawaii respectively in 1992, and the Kobe Earthquake in 1995, the structural engineers are again reminded to re-evaluate and further study the important issues of structural safety and reliability through structural control in preparation of such unavoidable environmental hazards.

Many activities were initiated and followed through in response to such needs. For example, a US Panel on Structural Control Research was established under the auspices of NSF in 1989. The panel's organization consists of an Executive Committee with Professor George W. Housner as its chairman, and seven Working Groups. In the meantime, a Japanese counterpart panel on Seismic Control Research was also established by the Science Council of Japan and Professor Takuji Kobori is the chairman of the panel. Both the US and Japan panels have been working closely in the development of a joint US-Japan research agenda in the area of active and hybrid control.[1] As a cooperative efforts of these two panels, the International Association for Structural Control was formed with the objective of promoting coordination and communication on structural control. The association consists of a Board of Directors with Professor Housner as President and Professor Kobori as Vice President. Subsequently, the First World Conference on Structural Control was held in Pasadena, California in August 1994. There were 337 participants from 15 countries at the Conference.[2]

In the following sections, a review of some of the control algorithms and devices developed will be provided.

3.1.1. Review of Control Algorithms

The idea of active control for civil engineering structures started to emerge about two to three decades ago. For example, Zuk[3] discussed the concept of kinetic structures and Nordell[4] studied the use of active systems to resist any exceptionally high overloading of a given structure. Murata and Ito[5] investigated the possibility of using a gyroscope to reduce the response of suspension bridges and to increase their flutter speeds. Yao et al.[6,7] studied the use of initially slack cables in forming bilinear structures which were subjected to earthquake loads.

The concept of structural control was more formally presented to the structural engineering profession in 1971 by Yao.[8] As an example, the use of thruster engines to generate impulsive control forces was mentioned. Since the advocacy by Yao and others, a proliferation of literatures in this area using modern control theories started to appear, such as Yao et al.[9,10], Martin and Soong[11], Yang et al.[12,13], Roorda[14] and others just to name a few. In general, an active control system consists of: (i) sensors installed at suitable locations of the structure to measure either external disturbances or system responses or both; (ii) devices to process the measured information and to compute necessary control forces based on a given control algorithm; and (iii) actuators to produce the required active control forces. The active control forces are determined by the measured information provided from sensors as well as the particular control algorithm used. Depending on the use of measured information, the control systems are classified as closed loop, open loop, and closed-open loop control.

For closed loop control, only system responses were monitored and used to determine control forces.[14−24] For open loop control, only disturbances were monitored and used to determine control forces.[25−27] For open-closed loop control, both system responses and disturbances were monitored and used to determine the control forces.[25,27,28]

Many research efforts in active structural control have focused on developing control algorithms based on different control design criteria. Based on the types of control rule used, control algorithms are classified as:

Classical optimal control

The control forces are obtained by minimizing a cost function or performance index. Minimizing quadratic cost functionals leads to a Riccati type equation. Such solutions require a large amount of computational effort. However, for the elastic systems with stationary disturbances, the control gain matrix is time-independent. Thus, the Riccati equation can be solved off-line and the computational efforts are significantly reduced. A list of 270 references on optimal control algorithms can be found in a survey article by Robinson.[29]

Instantaneous optimal control

It has been pointed out that, since the external excitation is ignored in the derivation of the Riccati equation for classical optimal control algorithms, the classical optimal control is not truly optimal. A new control algorithm, instantaneous optimal control, which circumvents this difficulty has been established.[30,31] The basic idea is to consider the cost functional as time-dependent. Evaluations of the instantaneous optimal control algorithm have been carried out, both analytically and experimentally, for structures subjected to earthquake-type excitations.[30-33] The results indicate that the application of active control forces can indeed be effective in reducing structural vibration and preventing potential structural damage due to the action of the environmental loads.

Pole assignment

Pole assignment algorithms have been studied extensively and are given in the general control textbook such as[34]. Their application to the civil engineering structures is fruitful when only a few vibration modes contribute significantly to the structural responses.[11,35] Successful application of these algorithms requires judicious placement of sensors and actuators.

Independent modal space control

It is well known that the equations of motion of an n-DOF structural system can be decomposed into a set of n decoupled single DOF systems in the modal space by coordinate transformation. Control algorithms based on this coordinate transformation concept have been referred to as independent modal space control.[13,16,36−48] The method essentially shifts the problem of control design from a coupled $2n$-order structural system to n second-order uncoupled systems, a considerably simpler problem with substantial saving in computational effort. It is particularly attractive when only a few critical modes have to be controlled. Independent modal space control has been applied for a number of control design problems including civil engineering structures.[44,46,47,49−53]

Bounded state control

Active control algorithms designed to limit the state variables within pre-scribed bounds are of practical importance when applied to civil engineering structures. All pulse control strategies proposed in the literature fall into this category.[26,32,54−61] The pulse control procedures are relatively simple when compared with other modern control techniques. They require less on-line computation and are suitable to be used for the control of inelastic structures. Another approach of bounded state control is suggested by Lee and Kozin[62,63] using linear state feedback laws. However, their approach requires that the external excitations be within a given bounds.

In the above literature survey, it appears that in most of the studies of active control for civil engineering structures, only linear, deterministic structural systems subjected to deterministic loads were considered. However, envi-ronmental loadings have to be considered as stochastic in nature and the theory of random vibration has to be used. The stochastic responses of the structures with active control systems subjected to either stationary random excitations, such as the wind gusts, or nonstationary random excitations, such as the seismic ground acceleration, have been investigated.[12,24,30,32,37,64−69]

Some historical accounts of the formulation and development of active control research for civil engineering structures are available in survey articles, such as those by, Reinhorn and Manolis[70], Soong[71], and Yang and Soong.[72] A comprehensive state-of-the-art book on the theory and practice of active control has been completed by Soong.[73]

3.1.2. Review of Control Devices

3.1.2.1. Passive control devices

Since no input energy is needed, passive control devices have naturally attracted considerable attention. Some passive control devices are reviewed as follows:

Tuned mass damper (TMD)

The concept of using a mass damper as a frequency-shifting and energy-absorbing system has been fruitful in the past.[74] Luft[75] and Warburton and Ayorinde[76] presented approximate formulas. for the optimal parametric design of a TMD system. An example of such technology is the use of a TMD system consisting of two 20-ton masses to reduce the second and fourth modes of vibration of the CN Tower in Toronto in order to stabilize the television antenna at the top.[77] A 400-ton and two 300-ton TMD systems have been installed on tops of the Citicorp Center in New York and the John Hancock Tower in Boston, respectively, to alleviate the structural vibration induced by the wind.[78,79] Recently, an efficient and compact TMD system was developed and used on the Iwakurojima-Bridge to reduce the wind-induced vibration of its pylons.[80]

Tuned liquid damper (TLD)

A tuned liquid damper is a device using the sloshing resonance of liquid in a tank to suppress the horizontal vibration of a structure. The advantages of TLD are that it is relatively easy to be tuned to the natural frequency of the structure and the cost of its construction and maintenance is low. Some methods have been proposed to tune the sloshing frequency, such as using rubber-covered surface[81] and submerged nets.[82,83] The TLD can be of different forms, such as tuned liquid column damper (TLCD)[84] and two-directional liquid column vibration absorber (LCVA).[85] The TLD system has already been installed in some high-rise buildings and bridges, such as the Shin Yokohama Prince Hotel, the Yokohama Marine Tower, the Sakitama bridge, and the Ikuchi bridge.[86,87]

Viscoelastic damper

A viscoelastic damper acts partly as a viscous material in which it functions as an energy absorber/dissipator, and partly as an elastic material, in which it functions as an energy storer. In general, there are two types of viscoelastic treatment: those in which the deformation is predominantly tensional, and those in which shear motions are dominant. The first type is achieved by direct application of a viscoelastic material to the surface of a structure in either a constrained or free layer. The second type of damper is one in which nearly all the deformation is in shear. It has been shown that shear damping is generally more effective than tensional deformation and is more suitable where large amounts of energy must be dissipated.[88] The development of the viscoelastic damper is not limited to numerical analysis or laboratory experiment only, there have already been some practical applications. One of them is the World Trade Center in New York with 20,000 dampers, the other is the Columbia Center in Seattle with 260 dampers. The results demonstrate that the viscoelastic damper is quite effective in reducing the structural responses due to both wind and earthquake loads.[89,90]

Other energy dissipated dampers

Several other types of energy dissipated dampers have been proposed and tested. Among the earlier examples are the frictional damping system[91-93] and the U-shaped mild steel damper.[94] The Kajima Corporation of Japan has developed two energy dissipated damping systems, the Honeycomb Damper system[95] and the Joint Damper system.[96] The Honeycomb Damper system has wide applicability and provides an excellent damping effect for both low-rise and high-rise buildings. It is currently applied in two medium-rise buildings, the Head Office Building of Oji Paper Co., Ltd. and the Sea Fort Square, in Tokyo. The Joint Damper system on the other hand can be used to reduce the adjacent building's sway. The first application of this system is the KI (Kajima Intelligent) Building in 1989 by the Kajima Corporation.

Base isolator

The idea of a base isolation system is to decouple a structure from the ground and protect it from the damaging effects of earthquake motion.[97] Several types of base isolator have been developed and tested, such as the resilient friction base isolator[98], the pure-friction base isolator[99], and the laminated rubber bearing.[100] Although the base isolator has the benefits of reducing acceleration and inter-story drift[101], the application is limited to low-rise rigid structures because of the possibility of generating uplift forces in the isolators for tall buildings.[102]

Active control devices

The active control devices are reviewed as follows:

Active mass driver (AMD)

The idea of using an AMD to mitigate structural response is to provide control forces by means of an electrical or electro-hydraulic type of actuator acting between a structure and an auxiliary mass.[103] Numerical analyses and laboratory verifications on the AMD system have attracted considerable attention.[104-106] The practical application of AMD did not come into reality until the construction of the Kyobashi Seiwa Building in August 1989.[107−109] Two AMD systems have been installed on the top floor of this 11-story office building in Tokyo to control lateral and torsional vibration, respectively. It is demonstrated that by activating the AMD systems, the response of the building due to both earthquake and strong wind loads can be suppressed to a very satisfactory degree. Since then, the AMD system has found its application in some other high-rise buildings in Japan, such as, the ORC200 Symbol Tower (188 m) in Osaka and the Land Mark Tower (282 m) in Yokohama.[86]

Active tendon

The active tendon system generally consists of a set of pre-stressed tendons connected between the electro-hydraulic servo-mechanisms and the joints placed at suitable locations in the structures. The control forces in the tendons are produced by the movement of hydraulic rams. Studies of such control systems have been done both analytically[14−17,36−39,54,55,64,65,110] and experimentally.[18,32,33,111−115] Full scale field implementation of an active tendon system was reported in 1991, which involved installing four actuators at the first floor of a two-bay six-story building located in Tokyo. Two modified control algorithms based on the classical optimal closed-loop control theory were proposed and tested. The results demonstrated the effectiveness of the active tendon system.[116]

Aerodynamic appendages

The use of aerodynamic appendages as an active control device to reduce the vibration of tall flexible buildings under strong winds was proposed by Klein et al.[117] The main attractive feature is that the control designer is able to exploit the energy in the wind to control the structure, which is being excited by the same wind. Thus, it is not necessary to provide external energy to produce control forces. Such an active control system has been continuously studied both analytically[20,21,118] and experimentally.[22]

Pulse generator

Gas pulse generators, producing pulsed forces generated by release of air jets, have been proposed for implementation of pulse control algorithms.[56,57] The corresponding magnitudes of the pulse forces are computed by the control law and response measurements. Such a control system has also been continuously investigated both analytically[26,58−60,119,120] and experimentally.[121,122]

Active variable stiffness (AVS)

The active variable stiffness system is a nonresonant type of seismic response control system developed by the Kajima Corporation of Japan[123] to reduce the response of buildings. The AVS system operates by actively controlling the stiffness of structures so as to prevent the structure from resonating with the earthquake motion.[107] The special characteristics of this AVS system are that, it can be accommodated behind a ceiling, therefore it is not necessary to reserve any particular space for installation of the system. Also, it has been designed as an energy-saving type of system and can be operated using only a small amount of electric power. The AVS system has been intensively tested in the laboratory.[124,125] The first application of the AVS system was reported from Japan in October, 1990[125], where the AVS system was installed on a three-story steel building (Central Building for new large-scale shaking table at the Kajima Technical Research Institute). The subsequent monitoring of this building demonstrated that the AVS system is quite effective in reducing the structural response due to earthquake.

Other active control devices

Other active control devices have also been proposed, such as gyroscopes[5], adaptive members and smart materials.[126] Most of research efforts on these devices have been focused on aerospace applications, such as advanced aircraft, launch vehicles, and large space-based platforms. Recently, a gyroscope stabilizer which made use of precession motion of a gyro to suppress the bending deflection of a tower-like structure was proposed by the Taisei Corporation of Japan.[127]

Hybrid control devices

As an alternative to pure passive or active control, the concept of hybrid structural control utilizing both active and passive control systems in an optimal fashion has also been proposed. The philosophy of hybrid structural control is to retain the beneficial characteristics in each individual control device and, in the meantime, compensate or cancel out the undesirable effects. Some of the hybrid control system which have been developed or currently under study are reviewed as follows:

Active tuned mass damper (ATMD)

In attempts to reduce the mass of the TMD system and limit the stroke length or excursion associated with the TMD while keeping building motion within acceptable limits, several researchers have proposed adding an actuator to the TMD system.[37,128−130] Experimental development on the ATMD system has since gained significant progress.[131,132] Tanida *et al.* proposed and developed a two-axis ATMD system which included the passive mechanism of a pendulum by the use of a curve-mass, so that the actuator requires much less power than does the conventional active mass damper.[132,133]

Active base isolator

The concept of active isolation as an alternative approach in structural design for earthquake hazard mitigation has been presented in[61,134−138] The idea involves installing an actuator at the base floor of a base-isolated building. The equations of motion are set up in the inertia coordinate system, thus the ground excitation input is introduced only at the base floor level. It is demonstrated that the control scheme designed using the active isolation concept for minimizing absolute motion of base-isolated structures under seismic excitation is powerful for absolute motion reduction.[138]

Other hybrid system

Other hybrid control systems utilizing both active and passive control devices have been reported but mainly in numerical studies.[139−143] Two seismic control system, using base isolators, TMD, and AMD, for protecting building structures against strong earthquakes have been proposed by Yang *et al.*[139] The results demonstrate that a reduction of more than 50% for the first story deflection can easily be accomplished using the hybrid control systems. Extension of the hybrid system to the control of nonlinear and hysteric structures has also been studied.[140] Another study on hybrid control, using base isolators and actuators, of a two and eight story lumped-mass shear-beam model was also conducted by Yang *et al.*[141]

As a final note on the applications of control devices in the civil engineering industry, it is worth mentioning that there have been thirty-six applications on buildings[86] and thirty-seven applications on bridges[87] in Japan along up to 1993.

3.1.3. Overview of the Present Research and Objectives

It has been demonstrated that for structures subjected to seismic loads, the Riccati closed-loop control law does not satisfy the classical optimal condition. Moreover, the optimal open-loop and closed-open-loop control algorithms are not applicable, because the earthquake ground motion is not known *a priori*.[30] The instantaneous optimal control techniques, developed by Yang and his colleagues[30,139,140], were proposed specifically for earthquake excited structures.

The basic idea in the derivation of instantaneous optimal control algorithm is to assume a time-dependent quadratic objective function. By minimizing the objective function at every time instant, three different types of instantaneous optimal control techniques, namely, closed-loop, open-loop, and closed-open-loop can be derived. Evaluations of the instantaneous optimal control algorithm have also been carried out experimentally for structures subjected to earthquake-type excitation.[32] It has been pointed out[30] that one of the advantages of using instantaneous optimal control algorithm is that it eliminates the need to solve the time-consuming and cumbersome Riccati matrix equation. This is particularly evident when the number of degrees of freedom of the structure is large.[32] However, the derivation of the algorithm involves time discretization of the equations of motion. Such discretization does not seem to guarantee the stability of the controlled structure. Yang *et al.*[142,143] proposed a systematic way of assigning the weighting matrix by using the Lyapunov direct method to ensure the stability of the controlled structures. The solution nevertheless involves solving the Riccati matrix equation.

In view of the current numerical studies of control applications in civil engineering available in the literature, most of the structures have been modeled using the simple lumped-mass shear-beam model. The main advantage of this approach is in its simplicity and numerical efficiency because only one degree of freedom is assigned to each floor. However, the shear-beam model is based on the assumption that the floor beams are perfectly rigid.[144] It has been shown that significant error in natural frequencies of the frame occurs when the floor beams are not much stiffer than the columns.[145] Moreover, the shear-beam model does not provide detailed information such as the distribution of displacements and stresses in the building frame which is of interest to the structural designers. To complement the simple shear-beam model, finite element modeling provides additional flexibility and accuracy as well as detailed knowledge of displacements and stresses. Nowadays, many general purpose finite element programs are available commercially, such as, ANSYS, MARC, ADINA, ABAQUS, and NONSAP, etc. Some implicit numerical integration schemes, such as, Newmark, Wilson-θ, and Houbolt methods have been implemented in the programs for dynamic transient analysis. These schemes are usually coded into the programs in the second order form which differs from the state-space form conventionally used in the derivation of optimal control algorithms. Because of this incompatibility, the incorporation of the control algorithms in the state-space form into the available commercial finite element programs is not routine. It appears that for the sake of easy implementation of the control algorithm, there is an advantage not to use state-space formulation.

In the present study, we propose two techniques which can be used on the control of civil engineering structures. First, we derive a form of closed-loop complete feedback control algorithm for the control of a structure modeled as a single degree-of-freedom (SDOF) system using an active tuned mass damper. The SDOF system is assumed to be under stationary Gaussian white noise ground excitation. The control force is calculated from the acceleration, velocity and displacement feedbacks of the structure and the active mass damper. The passive properties, including the stiffness and damping constants of the tuned mass damper, as well as the gain coefficients of the actuator are derived by minimizing the displacement variance of the SDOF system. The stability of the proposed algorithm is also discussed using the Routh-Hurwitz criterion. Numerical simulations are performed to evaluate the efficiency of the active tuned mass damper design.

Secondly, we develop a control algorithm based on the instantaneous optimal closed-loop control technique for the control of buildings subjected to earthquake and wind loadings, respectively. It is proposed that the building frames be modeled using finite elements, with a two-node six-degree-of-freedom planar frame element as a basic tool for the example analysis. The instantaneous optimal closed-loop control algorithm is derived in the second order form using the Newmark integration scheme. The weighting matrices are chosen so that the instantaneous control technique is unconditionally stable. Examples of optimal control of a three-bay ten-story steel building frame are presented using base isolators and active tendons. The input excitations are chosen based on the NS component of the ground acceleration record of the 1940 El Centro earthquake and some simulated wind loadings. Results on reduction of the dynamic response and the control effectiveness of the passive, active, and hybrid designs are presented and discussed.

3.2. CONTROL EFFICIENCY OF A SINGLE-DEGREE-OF-FREEDOM SYSTEM

It is well-known that the effect of applying a feedback control force to a structure is equivalent to modifying the damping and stiffness characteristics of the structure. In this section, the control efficiency of a single-degree-of-freedom system is discussed from the random vibration perspective.

The equation of motion for a single-degree-of-freedom (SDOF) system subjected to an external disturbance force $f(t)$ and a control force $u(t)$ can be written as (see Figure 1)

$$m\ddot{x}(t) + c\dot{x}(t) + kx(t) = f(t) + u(t), \tag{1}$$

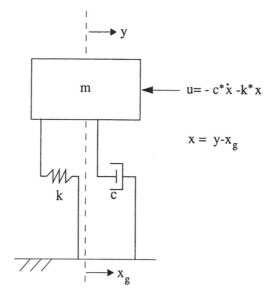

Figure 1. A single-degree-of-freedom system with closed-loop feedback control force under stationary Gaussian white noise ground excitation.

where m, c, and k represent the mass, damping, and stiffness coefficients, and $x(t)$, $\dot{x}(t)$ and $\ddot{x}(t)$ are the displacement, velocity, and acceleration of the SDOF system relative to the ground, respectively. For more general multiple degree-of-freedom systems which we will discuss later, the equations of motion can similarly be described as in (1).

Suppose that the control force $u(t)$ is governed by the response of the SDOF system and is given by

$$u(t) = -k^*x(t) - c^*\dot{x}(t), \tag{2}$$

where k^* and c^* are the control gains which also represent the equivalent stiffness and damping coefficients induced by the control force, respectively. Substituting (2) into (1) gives a controlled equation of motion for the SDOF system as

$$m\ddot{x}(t) + c(1 + r_c)\dot{x}(t) + k(1 + r_k)x(t) = f(t) \tag{3}$$

with the nondimensionalized control gains r_c and r_k defined as

$$r_c = \frac{c^*}{c}; r_k = \frac{k^*}{k}. \tag{4a,b}$$

Suppose that the SDOF system is subjected to a ground excitation $\ddot{x}_g$, then

$$f(t) = -m\ddot{x}_g \tag{5}$$

Furthermore, assuming that $\ddot{x}_g$ is a zero-mean stationary Gaussian white noise random process with a constant spectral density S_0, the variances of the relative displacement σ_x^2, relative velocity $\sigma_{\dot{x}}^2$ and absolute acceleration $\sigma_{\ddot{y}}^2$; where $\ddot{y} = \ddot{x} + \ddot{x}_g$, of the SDOF system, and the variance of the control force σ_u^2 are given as[146]

$$\sigma_x^2 = \frac{\pi m^2 S_0}{ck(1 + r_c)(1 + r_k)}, \tag{6a}$$

$$\sigma_{\dot{x}}^2 = \frac{\pi m S_0}{c(1 + r_c)}, \tag{6b}$$

$$\sigma_{\ddot{y}}^2 = \frac{\pi k S_0}{c}\left[\frac{1 + r_k}{1 + r_c} + 4\xi^2(1 + r_c)\right], \tag{6c}$$

$$\sigma_u^2 = \frac{\pi k m^2 S_0}{c}\left[\frac{r_k^2}{(1 + r_c)(1 + r_k)} + 4\xi^2\frac{r_c^2}{1 + r_c}\right], \tag{6d}$$

where

$$\xi = \frac{c}{2\sqrt{km}} \tag{7}$$

is the damping ratio of the SDOF system.

Displacement Feedback ($r_c = 0$)

In the case of displacement feedback, the nondimensionalized variances can be derived from (6) as follows,

$$\frac{ck}{\pi m^2 S_0}(\sigma_x^2)_D = \frac{1}{1 + r_k}, \tag{8a}$$

$$\frac{c}{\pi m S_0}(\sigma_{\dot{x}}^2)_D = 1, \tag{8b}$$

$$\frac{c}{\pi k S_0}(\sigma_{\ddot{y}}^2)_D = 1 + r_k + 4\xi^2, \tag{8c}$$

$$\frac{c}{\pi k m^2 S_0}(\sigma_u^2)_D = \frac{r_k^2}{1 + r_k}, \tag{8d}$$

These four equations are plotted in Figure 2 with $\xi = 0.05$.

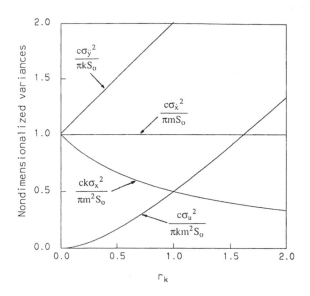

Figure 2. Nondimensionalized variances of relative displacement, relative velocity, absolute acceleration, and control force as functions of the nondimensionalized displacement feedback gain r_k for the case of displacement feedback ($\xi = 0.05$).

Velocity Feedback ($r_k = 0$)

Similarly, in the case of velocity feedback, the nondimensionalized variances can be derived from (6) as follows,

$$\frac{ck}{\pi m^2 S_0} (\sigma_x^2)_v = \frac{1}{1 + r_c}, \tag{9a}$$

$$\frac{c}{\pi m S_0} (\sigma_{\dot{x}}^2)_v = \frac{1}{1 + r_c}, \tag{9b}$$

$$\frac{c}{\pi k S_0} (\sigma_{\ddot{y}}^2)_v = \frac{1}{1 + r_c} + 4\xi^2(1 + r_c), \tag{9c}$$

$$\frac{c}{\pi k m^2 S_0} (\sigma_u^2)_v = 4\xi^2 \frac{r_c^2}{1 + r_c}, \tag{9d}$$

These four equations are plotted in Figure 3 with $\xi = 0.05$.

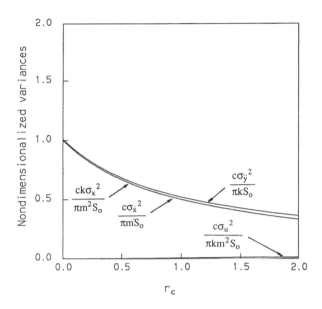

Figure 3. Nondimensionalized variances of relative displacement, relative velocity, absolute acceleration, and control force as functions of the nondimensionalized velocity feedback gain r_c for the case of velocity feedback ($\xi = 0.05$).

To further evaluate and compare the control efficiency using displacement and velocity feedback, the results of the nondimensionalized variances of the relative displacement, relative velocity, absolute acceleration are plotted against the nondimensionalized variance of the control force in Figures 4, 5 and 6, respectively. It can be seen that in all three figures, the results using velocity feedback ($r_k = 0$) all fall below those using displacement feedback ($r_c = 0$) for the ranges of damping ratios considered.

Monte Carlo simulations are used to verify the analytic derivations. Figures 7 and 8 show the results of the nondimensionalized variances of the relative displacement and absolute acceleration of the SDOF system with $\xi = 0.05$ plotted against the nondimensionalized variance of the control force, respectively. Three cases are considered: $r_k = 0$ (velocity feedback), $r_c = 0$ (displacement feedback), and $r_c = r_k$ (velocity and displacement feedback). Fairly good agreement between the results using Monte Carlo simulation and those of the analytic solutions are seen.

As a final note, it is seen from (8a) and (9a) that the variances of the relative displacement calculated using displacement feedback and velocity feedback are equal when $r_c = r_k$. Substituting this condition into (8d)

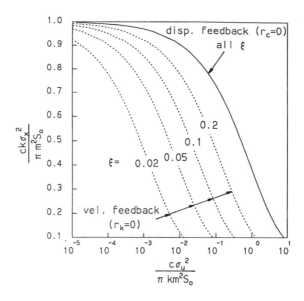

Figure 4. Nondimensionalized variance of relative displacement versus nondimensionalized variance of control force for the cases of displacement and velocity feedback.

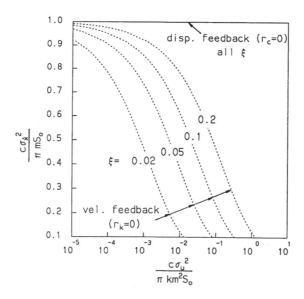

Figure 5. Nondimensionalized variance of relative velocity versus nondimensionalized variance of control force for the cases of displacement and velocity feedback.

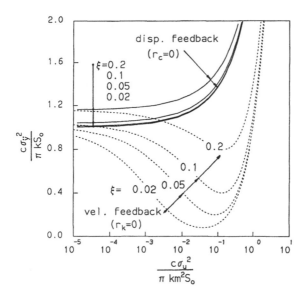

Figure 6. Nondimensionalized variance of absolute acceleration versus nondimensionalized variance of control force for the cases of displacement and velocity feedback.

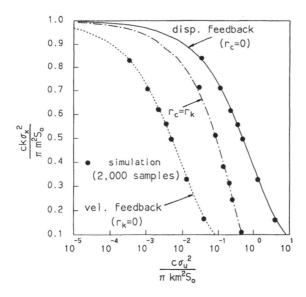

Figure 7. Monte Carlo simulations of the nondimensionalized variances of relative displacement and control force for the SDOF system with $\xi = 0.05$.

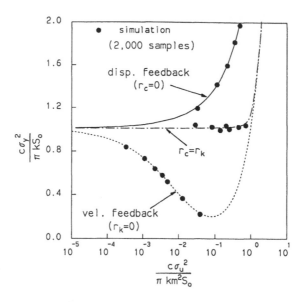

Figure 8. Monte Carlo simulations of the nondimensionalized variances of absolute acceleration and control force for the SDOF system with $\xi = 0.05$.

and (9d), the ratio of the control variances required using displacement feedback and velocity feedback can be found as

$$\frac{(\sigma_u^2)_D}{(\sigma_u^2)_v} = \frac{1}{4\xi^2}.$$

(10)

It can be concluded from (10) that when the damping ratio ξ is smaller than 0.5, to achieve the same amount of reduction in the variance of the relative displacement of the SDOF system, the variance of the control force is always smaller using velocity feedback. In other words, the velocity feedback is more efficient than the displacement feedback.

3.3. CONTROL OF BUILDINGS USING ACTIVE TUNED MASS DAMPER

In this section, a closed-loop complete feedback control algorithm is proposed for the control of structures modeled as a single degree-of-freedom (SDOF) system using an active tuned mass damper (ATMD). The SDOF system is assumed to be under stationary Gaussian white noise ground excitation. The control force is calculated from the acceleration, velocity and displacement feedbacks of the SDOF system and the active mass damper. The stiffness and damping coefficients of the tuned mass damper and the gain coefficients of the actuator are derived by minimizing the displacement variance of the SDOF system. The stability of the proposed algorithm is also discussed using the Routh-Hurwitz criterion. Simulations are performed to evaluate the efficiency of the ATMD design on a ten-story, three-bay building frame.

3.3.1. Problem Formulation

Assume that the equations of motion for a building structure can be described by the following second order equation,

$$M\ddot{x}(t) + C\dot{x}(t) + Kx(t) = F(t) + Bu(t), \tag{11}$$

where $x(t) \in \Re^n$ is the displacement vector (relative to the ground), $u(t) \in \Re^m$ is the control force vector; n and m represent the total number of degrees of freedom and the number of actuators, respectively. The constant matrices M, C, and K denote the $n \times n$ structural mass, damping, and stiffness matrices, respectively. The matrix B indicates the location of actuators and the vector $F(t)$ denotes the external disturbance. Note that (11) is a multiple dimensional representation of (1) considered in Section 3.2. Assuming the structure is subjected to a ground excitation $\ddot{x}_g(t)$, the external force $F(t)$ can be expressed as

$$F(t) = -Mr\ddot{x}_g(t), \tag{12}$$

where the vector $r \in \Re^n$ indicates the location of the excitation.

For a linear system with proportional damping, the equations of motion can be decoupled in the normal coordinates. Since the dynamic behavior of a structure is usually governed by a dominant mode, it is logical to design one controller specifically for this mode. The equation of motion of the dominant mode, which is a SDOF system, can be written as

$$m_0\ddot{x}_0(t) + c_0\dot{x}_0(t) + k_0x_0(t) = \beta_0m_0\ddot{x}_g(t) - u(t) \tag{13}$$

where $x_0(t) \in \Re$ represents the modal amplitude. The definitions of the modal mass, damping, and stiffness coefficients m_0, c_0, and k_0, and the modal participation factor β_0, can be found, for example, in the book by Clough and Penzien.[144] It is noted that in (13), the eigenvector of the dominate mode Φ_0 is scaled proportionally according to the relationship

$$\Phi_0'B = 1. \tag{14}$$

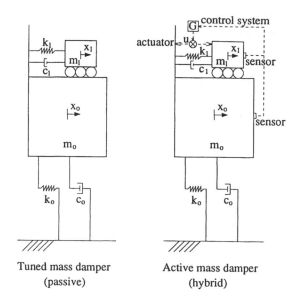

Figure 9. Modelings of the passive and hybrid types of control system using an auxiliary mass.

The modal amplitude x_0 is related to the displacement vector x by the following equation

$$x_0 = \frac{\Phi_0' M}{\Phi_0' M \Phi_0} x. \tag{15}$$

If an auxiliary mass damper is added onto the SDOF system described in (13) and also a control force is acting between the SDOF system and the auxiliary mass (see Figure 9), then the resulting two degree-of-freedom system can be written as

$$\begin{bmatrix} m_0 & 0 \\ 0 & m_1 \end{bmatrix} \begin{Bmatrix} \ddot{x}_0 \\ \ddot{x}_1 \end{Bmatrix} + \begin{bmatrix} c_0 + c_1 & -c_1 \\ -c_1 & c_1 \end{bmatrix} \begin{Bmatrix} \dot{x}_0 \\ \dot{x}_1 \end{Bmatrix}$$
$$+ \begin{bmatrix} k_0 + k_1 & -k_1 \\ -k_1 & k_1 \end{bmatrix} \begin{Bmatrix} x_0 \\ x_1 \end{Bmatrix} = -\begin{Bmatrix} \beta_0 m_0 \\ m_1 \end{Bmatrix} \ddot{x}_g + \begin{Bmatrix} -1 \\ 1 \end{Bmatrix} u \tag{16}$$

where m_1, c_1 and k_1 are the mass, damping and stiffness coefficients of the auxiliary mass damper, x_0 and x_1 represent the displacements of the SDOF system and the mass damper relative to the ground, respectively.

3.3.2. Tuned Mass Damper (TMD)

When the control force is inactive, the auxiliary mass damper is of passive type and attenuation of the response of the SDOF system is due solely to the mass damper parameters, m_1, c_1 and k_1. By properly adjusting these parameters, the response of the SDOF system can be reduced to various degrees. Ayorinde and Warburton[76,147] proposed a method of determining c_1 and k_1 as functions of a predetermined m_1 by minimizing the response of the SDOF system with the participation factor $\beta_0 = 1$. The derivation is extended in the following for the case with arbitrary β_0 value. Assuming that the ground excitation $\ddot{x}_g(t)$ is a stationary Gaussian white noise random process with spectral density S_0, the variance of the displacement of the SDOF system, $\sigma_{x_0}^2$, can be derived by assuming $\xi_0 = 0$ as[76]

$$
\sigma_{x_0}^2 = \frac{\pi S_0}{2\omega_0^3} \left\{ \frac{1}{\mu_1 f_1 \xi_1} \left[\beta_0^2 - (1+\mu_1)^2 (2\beta_0 - \mu_1) f_1^2 \right. \right.
$$
$$
\left. \left. + (1+\mu_1)^4 f_1^4 \right] + \frac{4\xi_1 f_1}{\mu_1} (1+\mu_1)^3 \right\} \tag{17}
$$

with

$$
\omega_i = \left(\frac{k_i}{m_i} \right)^{1/2} \; ; \; \xi_i = \frac{c_i}{2\sqrt{m_i k_i}} \; , \quad i = 0, 1, \tag{18a,b}
$$

$$
\mu_i = \frac{m_1}{m_0} \; ; \; f_1 = \frac{\omega_1}{\omega_0}. \tag{19,a,b}
$$

The optimal values of f_1 and ξ_1 can be found by taking derivative of $\sigma_{x_0}^2$ with respect to f_1 and ξ_1 and setting them equal to zero, respectively. Solving these two equations gives

$$
f_{1_{opt}} = \frac{\left(\alpha - \frac{1}{2}\mu_1 \right)^{1/2}}{1 + \mu_1} \; ; \; \xi_{1_{opt}} = \left[\frac{\mu_1 \left(\alpha - \frac{1}{4}\mu_1 \right)}{4(1 + \mu_1) \left(\alpha - \frac{1}{2}\mu_1 \right)} \right]^{1/2}, \tag{20a,b}
$$

where

$$
\alpha = \frac{\beta_0 + \beta_0 \mu_1}{\beta_0 + \mu_1} \tag{21}
$$

Now, substituting (20) into (18) and (19) yields

$$
k_{1_{opt}} = \frac{\mu_1 \left(\alpha - \frac{1}{2}\mu_1 \right)}{(1 + \mu_1)^2} k_0 \; ; \; c_{1_{opt}} = \frac{\mu_1^{3/2} \left(\alpha - \frac{1}{4}\mu_1 \right)^{1/2}}{2\xi_0 (1 + \mu_1)^{3/2}} c_0. \tag{22a,b}
$$

3.3.3. Active Tuned Mass Damper Using Velocity Feedback

In addition to adjusting the passive parameters m_1, c_1 and k_1, the response of the SDOF system can be further reduced by activating the actuator installed between the system and the auxiliary mass. The control force generated by the actuator can be designed by using either the optimal control theory[38,128] or the pole assignment method.[148] One of the optimal control algorithms which has been designed and implemented on a ten-story office building in Tokyo by the Kajima Corporation of Japan involves feeding back the velocity of both the SDOF system and the auxiliary mass to calculate the control force[149,150], i.e.

$$u(t) = g_0 \dot{x}_0(t) - g_1 \dot{x}_1(t) \tag{23}$$

where g_0 and g_1 are the velocity feedback gains. Substituting (23) into (16) and moving the terms associated with the control force to the left hand side of the equations give,

$$\begin{bmatrix} m_0 & 0 \\ 0 & m_1 \end{bmatrix} \begin{Bmatrix} \ddot{x}_0 \\ \ddot{x}_1 \end{Bmatrix} + \begin{bmatrix} c_0 + c_1 + g_0 & -c_1 - g_1 \\ -c_1 - g_0 & c_1 + g_1 \end{bmatrix} \begin{Bmatrix} \dot{x}_0 \\ \dot{x}_1 \end{Bmatrix}$$
$$+ \begin{bmatrix} k_0 + k_1 & -k_1 \\ -k_1 & k_1 \end{bmatrix} \begin{Bmatrix} x_0 \\ x_1 \end{Bmatrix} = - \begin{Bmatrix} \beta_0 m_0 \\ m_1 \end{Bmatrix} \ddot{x}_g. \tag{24}$$

Two interesting observations can be made from (24). First, the damping matrix becomes unsymmetrical if $g_0 \neq g_1$, thereby the system becomes potentially unstable. The stability of (24) can be verified using the Routh-Hurwitz stability criterion.[151,152] The characteristic Equation (24) can be written as

$$a_4 s^4 + a_3 s^3 + a_2 s^2 + a_1 s + a_0 = 0, \tag{25}$$

where

$$a_4 = m_0 m_1, \tag{26a}$$
$$a_3 = m_0 c_1 + m_0 g_1 + m_1 c_0 + m_1 c_1 + m_1 g_0, \tag{26b}$$
$$a_2 = m_0 k_1 + m_1 k_0 + m_1 k_1 + c_0 c_1 + c_0 g_1, \tag{26c}$$
$$a_1 = k_0 c_1 + k_0 g_1 + k_1 c_0, \tag{26d}$$
$$a_0 = k_0 k_1. \tag{26e}$$

The Routh-Hurwitz stability criterion states that the roots of the characteristic Equation (25) have negative real parts; in other words, the controlled system (24) is stable, if and only if the following conditions are satisfied,

$$a_0, a_1, a_2, a_3, a_4 > 0, \tag{27a}$$
$$a_2 a_3 - a_1 a_4 > 0, \tag{27b}$$
$$a_1 a_2 a_3 - a_1^2 a_4 - a_0 a_3^2 > 0. \tag{27c}$$

Neglecting the terms associated with c_0 and c_1[153,154] and substituting (26) into (27) give

$$g_0 > 0 \; ; \; g_1 > 0 \; ; \; g_1 > \frac{\mu_1 f_1^2}{1 - f_1^2} g_0. \qquad (28a,b,c)$$

For the case of optimally tuned mass damper where $k_1 = k_{1_{opt}}$ and $c_1 = c_{1_{opt}}$ as given in (22), the condition (28c) can be further simplified as follows,

$$g_1 > \frac{2\alpha\mu_1 - \mu_1^2}{2(1 - \alpha) + 5\mu_1 + 2\mu_1^2} g_0. \qquad (29)$$

As regards the second observation, if $g_0 = g_1$ in (24), the damping matrix remains symmetrical and positive definite. The equations of motion become the same as those of a passive tuned mass damper system as discussed in the previous section with $c_1 + g_0$ replacing c_1 in (16). Since the optimal values $k_{1_{opt}}$ and $c_{1_{opt}}$ derived in (22) minimize $\sigma_{x_0}^2$, the implication is that under the optimally tuned condition of $k_1 = k_{1_{opt}}$ and $c_1 = c_{1_{opt}}$, the addition of control force using either velocity or displacement feedback would not decrease the displacement variance of the SDOF system $\sigma_{x_0}^2$. To further reduce the response of this SDOF system, one of the choices would be to include a feedback consisting all three quantities, namely acceleration, velocity and displacement (henceforth referred to as complete feedback).

3.3.4. Active Tuned Mass Damper Using Complete Feedback

Assume that the control force u is chosen to be a function of the acceleration, velocity and displacement of the SDOF system and the auxiliary mass in the following form

$$u = m_2(\ddot{x}_0 - \ddot{x}_1) + c_2(\dot{x}_0 - \dot{x}_1) + k_2(x_0 - x_1), \qquad (30)$$

where the feedback gains m_2, c_2 and k_2 can be considered as the equivalent mass, damping and stiffness coefficients, respectively, of the active control force. Substituting (30) into (16) and rearranging the terms give,

$$\begin{bmatrix} m_0 + m_2 & -m_2 \\ -m_2 & m_1 + m_2 \end{bmatrix} \begin{Bmatrix} \ddot{x}_0 \\ \ddot{x}_1 \end{Bmatrix} + \begin{bmatrix} c_0 + c_1 + c_2 & -c_1 - c_2 \\ -c_1 - c_2 & c_1 + c_2 \end{bmatrix} \begin{Bmatrix} \dot{x}_0 \\ \dot{x}_1 \end{Bmatrix}$$
$$+ \begin{bmatrix} k_0 + k_1 + k_2 & -k_1 - k_2 \\ -k_1 - k_2 & k_1 + k_2 \end{bmatrix} \begin{Bmatrix} x_0 \\ x_1 \end{Bmatrix} = - \begin{Bmatrix} \beta_0 m_0 \\ m_1 \end{Bmatrix} \ddot{x}_g. \qquad (31)$$

To simplify the derivation, let x_1 be expressed as follows,

$$x_1 = z_1 + x_0, \qquad (32)$$

with z_1 denoting the relative displacement between the SDOF system and the auxiliary mass. Substituting (32) into (31) yields

$$
\begin{bmatrix} m_0 + m_1 & m_1 \\ m_1 & m_1 + m_2 \end{bmatrix} \begin{Bmatrix} \ddot{x}_0 \\ \ddot{z}_1 \end{Bmatrix} + \begin{bmatrix} c_0 & 0 \\ 0 & c_1 + c_2 \end{bmatrix} \begin{Bmatrix} \dot{x}_0 \\ \dot{z}_1 \end{Bmatrix}
$$
$$
+ \begin{bmatrix} k_0 & 0 \\ 0 & k_1 + k_2 \end{bmatrix} \begin{Bmatrix} x_0 \\ z_1 \end{Bmatrix} = - \begin{Bmatrix} \beta_0 m_0 + m_1 \\ m_1 \end{Bmatrix} \ddot{x}_g. \tag{33}
$$

The displacement variance of the SDOF system $\sigma_{x_0}^2$ can then be found using the results from Crandall and Mark[146] as

$$
\sigma_{x_0}^2 = \pi S_0
$$
$$
\frac{\dfrac{B_0^2}{A_0}(A_2 A_3 - A_1 A_4) + A_3(B_1^2 - 2B_0 B_2) + A_1(B_2^2 - 2B_1 B_3) + \dfrac{B_3^2}{A_4}}{(A_1 A_2 - A_0 A_3)}
$$
$$
\frac{}{A_1(A_2 A_3 - A_1 A_4) - A_0 A_3^2} \tag{34}
$$

where

$$
A_4 = m_0 m_1 + m_0 m_2 + m_1 m_2, \tag{35a}
$$
$$
A_3 = (m_0 + m_1)(c_1 + c_2) + (m_1 + m_2)c_0, \tag{35b}
$$
$$
A_2 = (m_0 + m_1)(k_1 + k_2) + (m_1 + m_2)k_0, \tag{35c}
$$
$$
A_1 = c_0(k_1 + k_2) + (c_1 + c_2)k_0, \tag{35d}
$$
$$
A_0 = k_0(k_1 + k_2), \tag{35e}
$$
$$
B_3 = 0, \tag{35f}
$$
$$
B_2 = -(\beta_0 m_0 m_1 + \beta_0 m_0 m_2 + m_1 m_2), \tag{35g}
$$
$$
B_1 = -(\beta_0 m_0 + m_1)(c_1 + c_2), \tag{35h}
$$
$$
B_0 = -(\beta_0 m_0 + m_1)(k_1 + k_2) \tag{35i}
$$

Again, the optimal values of c_2 and k_2 can be found by taking the derivative of $\sigma_{x_0}^2$ with respect to $c_1 + c_2$ and $k_1 + k_2$ and setting them equal to zero, respectively. Neglecting the terms associated with c_0 and solving the two equations yield

$$
(k_1 + k_2)_{opt} = \frac{\mu_1 \left(\alpha + \mu_2 + \frac{\mu_2}{\mu_1} - \frac{1}{2}\mu_1 \right)}{(1 + \mu_1)^2} k_0 \tag{36a}
$$

$$
(c_1 + c_2)_{opt} = \frac{\mu_1^{3/2} \left[\alpha \left(1 + \mu_2 + \frac{\mu_2}{\mu_1} \right) - \frac{1}{4}\mu_1 \right]^{1/2}}{2\xi_0 (1 + \mu_1)^{3/2}} c_0, \tag{36b}
$$

where the equivalent mass ratio μ_2 is defined as

$$\mu_2 = \frac{m_2}{m_0}. \tag{37}$$

If k_1 and c_1 are equal to $k_{1_{opt}}$ and $c_{1_{opt}}$, respectively, then

$$k_{2_{opt}} = \frac{\mu_2}{1 + \mu_1} k_0, \tag{38a}$$

$$c_{2_{opt}} = \frac{\mu_1^{3/2} \left\{ \left[\alpha \left(1 + \mu_2 + \frac{\mu_2}{\mu_1} \right) - \frac{1}{4} \mu_1 \right]^{1/2} - \left(\alpha - \frac{1}{4} \mu_1 \right)^{1/2} \right\}}{2 \xi_0 (1 + \mu_1)^{3/2}} c_0 \tag{38b}$$

Moreover, substituting (36) into (34) gives

$$\sigma_{x_0}^2 \approx \frac{2\pi S_0}{\omega_0^3} \frac{(1 + \mu_1)^{3/2}}{\mu_1^{1/2}} \left[\alpha \left(1 + \mu_2 + \frac{\mu_2}{\mu_1} \right) - \frac{1}{4} \mu_1 \right]^{1/2}. \tag{39}$$

Equation (39) suggests that the displacement variance of the SDOF system $\sigma_{x_0}^2$ can be reduced when μ_2 is smaller than zero. The stability condition of this complete feedback control algorithm can be determined using the Routh-Hurwitz stability criterion as shown in (27), but with A_4, A_3, A_2, A_1, and A_0 given in (35) replacing a_4, a_3, a_2, a_1, and a_0 in (26). The algorithm is found to be stable when

$$\mu_2 > \max \left\{ -\frac{\mu_1 \left(\alpha - \frac{\mu_1}{2} \right)}{(1 + \mu_1)}, -\frac{\mu_1 \left(\alpha - \frac{\mu_1}{4} \right)}{\alpha (1 + \mu_1)} \right\} \tag{40}$$

3.3.5. Numerical Results

To evaluate the designs of the tuned mass damper (TMD) and the active tuned mass damper (ATMD) presented in this section, the control performance of a ten-story, three-bay building frame is analyzed and discussed.

A ten-story, three-bay moment resistant steel building frame used by Naiem[155] in the Seismic Design Handbook is adopted to demonstrate the effects of passive and active TMDs on the vibration control of a structure

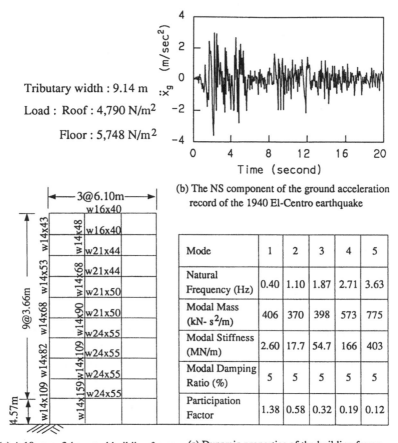

Tributary width : 9.14 m

Load : Roof : 4,790 N/m²

Floor : 5,748 N/m²

(b) The NS component of the ground acceleration record of the 1940 El-Centro earthquake

Mode	1	2	3	4	5
Natural Frequency (Hz)	0.40	1.10	1.87	2.71	3.63
Modal Mass (kN- s²/m)	406	370	398	573	775
Modal Stiffness (MN/m)	2.60	17.7	54.7	166	403
Modal Damping Ratio (%)	5	5	5	5	5
Participation Factor	1.38	0.58	0.32	0.19	0.12

(a) A 10-story 3-bay steel building frame   (c) Dynamic properties of the building frame

Figure 10. Modeling of a 10-story 3-bay steel building frame under the NS component of the ground acceleration record of the 1940 El Centro earthquake.

modeled as a multiple degrees of freedom system. The building frame is modeled using a two-node, six-degree-of-freedom planar beam-column element. One element is used to model each of the 70 members of the frame (30 beam and 40 column members). The design loads on the frame are assumed as, roof load = 4,790 N/m² and floor load = 5,748 N/m², all with tributary width 9.14 m. The total mass of this building frame is calculated as 963.5 KN-sec²/m. The damping ratios are assumed as 5% for each mode. The dynamic properties of the building frame, such as the natural frequency, modal mass, modal stiffness, and participation factor are calculated and listed in Figure 10. The first twenty seconds of the NS component of the ground acceleration record of the 1940 El Centro earthquake are used as the input excitation. It is noted that under the assumed building properties and the ground excitation, the displacement response due to the first mode constitutes approximately 90% of the total displacement response. Thus the first mode is selected for the designs of TMD and ATMD system in the following analyses.

In the following analyses, the time average quantities, such as the average displacement J_d, the average absolute acceleration $J_{\ddot{D}}$, and the average control force J_u are used for discussion. These values are calculated based on the equation

$$J_p = \frac{1}{t_f} \int_0^{t_f} p' p \, dt. \tag{41}$$

Tuned mass damper

It is assumed that a tuned mass damper with the stiffness of the damper tuned to the optimal value ($k_1 = k_{1_{opt}}$) is installed on top of the building frame. Three different mass ratio for the auxiliary mass; 1%, 2%, and 3% of the total mass of the building frame, respectively, are assumed for the study.

Figure 11 shows the average displacement J_d of the building frame for the damping ξ_1 ratio ranging between 0 to 20%. It is noted that significant decreases in J_d are seen as ξ_1 increases from 0 to 10% for all three values of mass ratio. For ξ_1 greater than 10%, however, slightly increases of J_d are seen. The solid dots in the figure represent the results of the tuned mass damper with optimal damping ratios calculated using (20b). It is seen that these dots seem to fall in the neighborhoods of the minimum J_d values for all three cases studied. This observation confirms that the set of optimal properties derived in (20) can minimize the response of the building frame. Figure 12 shows that the average absolute acceleration $J_{\ddot{D}}$ of the building frame as function of ξ_1. It is seen that $J_{\ddot{D}}$ decreases as ξ_1 increases for all three values of mass ratio.

The maximum floor displacements for the building frame controlled by the tuned mass damper with optimal properties are plotted in Figure 13. It is seen that the displacement reduction becomes more prominent as the mass ratio increases. Generally speaking, the reduction does not seem to be significant as only about 20% of the displacements are suppressed using a tuned mass damper weighted 3% of the total weight of the building. In order to reduce the seemingly heavy mass required for the TMD system, adding active control systems appears to be an alternative.

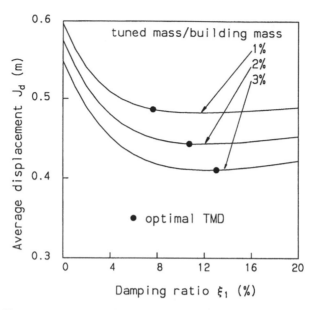

Figure 11. Average displacement J_d of the building frame controlled by a tuned mass damper with various values of damping ratio ξ_1.

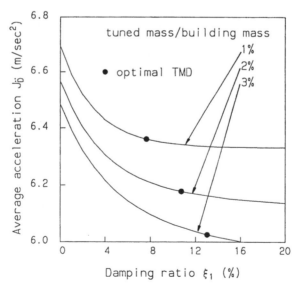

Figure 12. Average absolute acceleration $J_{\ddot{D}}$ of the building frame controlled by a tuned mass damper with various values of damping ratio ξ_1.

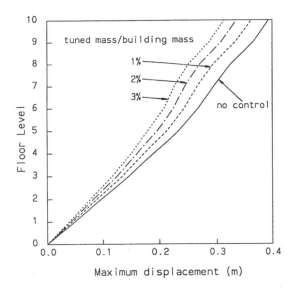

Figure 13. Maximum floor displacement profiles for the building frame controlled by a tuned mass damper with optimal properties.

Figure 14 shows the maximum absolute floor accelerations for the building frame. Although the results using the optimally tuned mass damper seem slightly smoother than those without control, the difference is not significant. Especially, the high acceleration value at the top floor level remains almost unchanged regardless of the presence of the tuned mass damper.

Active tuned mass damper using velocity feedback

Assume that an actuator is placed between the building and the TMD, and the control force is calculated based on the velocity feedback only. The mass of the damper is assumed to be 1% of the total mass of the building. The optimal frequency ratio $f_{1_{opt}}$ and damping ratio $\xi_{1_{opt}}$ are calculated as 0.974 and 0.0765, respectively. Two sets of properties are considered for the mass damper, one with $f_1 = 0.2$ and $\xi_1 = 0$, the other with $f_1 = f_{1_{opt}}$ and $\xi_1 = 0$. Also, the velocity feedback gain coefficient g_1 is set to be $0.4m_1\omega_1$, while the other coefficient g_0 varies from 0 to $0.1m_0\omega_0$. Figure 15 shows the results of the average displacement J_d and the average control force J_u using these two sets of properties. It is seen that for the TMD with $f_1 = 0.2$ and $\xi_1 = 0$, J_d decreases and J_u increases as $g_0/2m_0\omega_0$. varies from 0 to 0.05. In the meantime, both J_d and J_u increase for the TMD with optimal stiffness ratio.

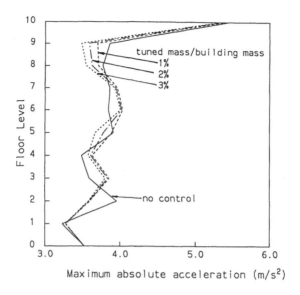

Figure 14. Maximum absolute floor acceleration profiles for the building frame controlled by a tuned mass damper with optimal properties.

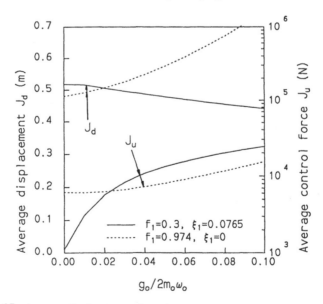

Figure 15. Average displacement J_d and average control force J_u of the building frame controlled by an active tuned mass damper with non-optimal properties using velocity feedback.

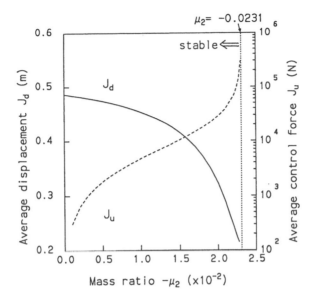

Figure 16. Average displacement J_d and average control force J_u as functions of equivalent mass ratio μ_2 for the building frame controlled by an active tuned mass damper using complete feedback.

Active tuned mass damper using complete feedback

Suppose that the control force generated by the actuator placed between the building and the TMD is calculated using complete feedback, and the mass ratio of the TMD is assumed to be 1% with the stiffness and damping tuned to their optimal values.

Figure 16 shows the results of the average displacement J_d and the average control force J_u plotted against the equivalent mass ratio μ_2. It is seen that J_d decreases and J_u increases as μ_2 varies from 0 to –0.023. This observation suggests that the optimally tuned ATMD with complete feedback is effective in reducing the displacement of the building frame. It is noted that, in this case, the complete feedback diverges when μ_2 is smaller than –0.0231. The divergence is due to the fact that the diagonal stiffness coefficient corresponding to the ATMD system becomes negative.

Figure 17 shows the maximum floor displacements of the building frame controlled by the ATMD using complete feedback with μ_2 equals to –0.01, –0.015, –0.02, and –0.022, respectively. The results for the frame without control and controlled by a TMD are also plotted in the figure for comparison. It is seen that the maximum floor displacement profile decreases steadily

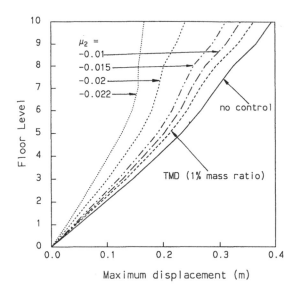

Figure 17. Maximum floor displacement profiles for the building frame controlled by an active tuned mass damper using complete feedback.

as μ_2 varies from –0.01 to –0.022. The largest control effect is seen when $\mu_2 = -0.022$, which gives approximately 60% reduction in displacement at the top of the building.

The results of the maximum absolute floor accelerations of the above six cases are plotted in Figure 18. Significant increases in the absolute acceleration are seen for the cases of $\mu_2 = -0.02$ and -0.022 at the sixth, seventh, and top floor levels. For the other four cases, only moderate variations are observed.

Comparisons

Figure 19 shows the comparison of the results between using complete feedback on a TMD with optimal properties and using velocity feedback on a TMD with non-optimal properties. The mass ratios for both TMDs are assumed as 1% of the total mass of the building frame. The properties of the non-optimal TMD are assumed as $f_1 = 0.2$ and $\xi_1 = 0$. This set of non-optimal properties is similar to those of an active mass driver system used by Kobori *et al.*[149] to suppress the response of a ten-story office building subjected to earthquake and typhoon. The results of the average

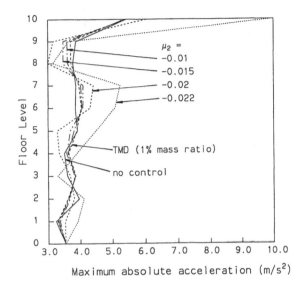

Figure 18. Maximum absolute floor acceleration profiles for the building frame controlled by an active tuned mass damper using complete feedback.

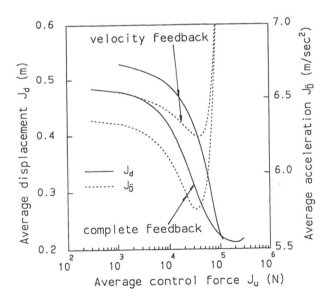

Figure 19. Comparisons of control efficiency between complete feedback and velocity feedback.

displacement J_d and the average absolute acceleration $J_{\ddot{D}}$ are plotted against the average control force J_u. It is seen that for both cases, as J_u increases, J_d decreases and $J_{\ddot{D}}$ decreases at first and increases after J_u exceeds 3×10^4 N. Between these two cases, it is observed that for the same amount of J_u, the average displacement and absolute acceleration of the building frame controlled by ATMD with optimal properties using complete feedback are smaller than those controlled by ATMD with non-optimal properties using velocity feedback for the range of J_u up to 7×10^4 N. This observation suggests that, for the cases studied and the parameters assumed, it is more efficient to control the building frame by ATMD with optimal properties using complete feedback.

3.4. INSTANTANEOUS OPTIMAL CONTROL ALGORITHMS

In this section, an application of instantaneous optimal control algorithm using the Newmark integration scheme is presented. The proposed control design procedure involves only simple algebraic manipulations and ensures the closed-loop stability by properly choosing the weighting matrices. To validate the current development, a ten-story three-bay building frame subjected to seismic loads and a tall building subjected to random wind loads are used as examples.

3.4.1. Problem Formulation

Note that the equations of motion for the building frame can be described as in Equation (11). The equations of motion (11) can be solved by a direct integration scheme known as the Newmark method. For the sake of completeness, the procedure of the Newmark method is briefly stated in the following:

$$x_t = x_{t-\Delta t} + \Delta x_t, \tag{42a}$$

$$\dot{x}_t = (1 - a_5)\dot{x}_{t-\Delta t} - a_6\ddot{x}_{t-\Delta t} + a_4\Delta x_t, \tag{42b}$$

$$\ddot{x}_t = (1 - a_3)\ddot{x}_{t-\Delta t} - a_2\dot{x}_{t-\Delta t} + a_1\Delta x_t, \tag{42c}$$

$$\Delta x_t = L^{-1}\beta F_t, \tag{42d}$$

$$L = a_1 M + a_4 C + K, \tag{42e}$$

$$\Delta F_t = F_t + Bu_t + M(a_2\dot{x}_{t-\Delta t} + a_3\ddot{x}_{t-\Delta t}) + C(a_5\dot{x}_{t-\Delta t} + a_6\ddot{x}_{t-\Delta t})$$
$$- (M\ddot{x}_{t-\Delta t} + C\dot{x}_{t-\Delta t} + Kx_{t-\Delta t}) \tag{42f}$$

with

$$a_1 = \frac{1}{\beta(\Delta t)^2}; a_2 = \frac{1}{\beta \Delta t}; a_3 = \frac{1}{2\beta}, \qquad \text{(43a,b,c)}$$

$$a_4 = \frac{\gamma}{\beta \Delta t}; a_5 = \frac{\gamma}{\beta}; a_6 = \Delta t(\frac{\gamma}{2\beta} - 1), \qquad \text{(43d,e,f)}$$

where γ and β are the Newmark parameters and, by choosing these parameters properly, the numerical convergence is assured unconditionally. The subscript "t" used in (42) represents the corresponding values at the time instant "t."

The objective of an instantaneous optimal control design is, given the previous structural response at time $t - \Delta t$, to find a displacement and velocity feedback of the form

$$u_t = G_1 x_t + G_2 \dot{x}_t, \qquad (44)$$

at time t; where $G_1, G_2 \in \Re^{m \times n}$ are constant gain matrices, so that the following time-dependent quadratic form performance index

$$J_t = \frac{1}{2}(x_t' Q_1 x_t + \dot{x}_t' Q_2 \dot{x}_t + u_t' R u_t), \qquad (45)$$

is instantaneously minimized at time t subject to (11). In (45), Q_1 and $Q_2 \in \Re^{n \times n}$ are positive semi-definite symmetric matrices and $R \in \Re^{m \times m}$ is a positive definite symmetric matrix. These matrices are the so-called *weighting matrices* and play an important role in the design of optimal control.

The instantaneous control input u_t which minimizes the performance index J_t at time instant t can be derived by first converting the constrained minimization problem to an unconstrained one by defining a Hamiltonian function as

$$\begin{aligned} H_t = {} & J_t + \lambda_1'(x_t - x_{t-\Delta t} - \Delta x_t) \\ & + \lambda_2'[\dot{x}_t - (1 - a_5)\dot{x}_{t-\Delta t} + a_6\ddot{x}_{t-\Delta t} - a_4\Delta x_t] \\ & + \lambda_3'[\ddot{x}_t - (1 - a_3)\ddot{x}_{t-\Delta t} + a_2\dot{x}_{t-\Delta t} - a_1\Delta x_t] \end{aligned} \qquad (46)$$

where λ_i; $i = 1, 2, 3$, are the Lagrange multipliers. The necessary conditions for minimizing the performance index J_t are given by

$$\frac{\partial H_t}{\partial x_t'} = \frac{\partial H_t}{\partial \dot{x}_t'} = \frac{\partial H_t}{\partial \ddot{x}_t'} = \frac{\partial H_t}{\partial u_t'} = \frac{\partial H_t}{\partial \lambda_1'} = \frac{\partial H_t}{\partial \lambda_2'} = \frac{\partial H_t}{\partial \lambda_3'} = 0 \qquad (47)$$

Substituting (46) into (47) yields

$$Q_1 x_t + \lambda_1 = 0, \tag{48a}$$

$$Q_2 \dot{x}_t + \lambda_2 = 0, \tag{48b}$$

$$\lambda_3 = 0, \tag{48c}$$

$$R u_t - B' L^{-1}(\lambda_1 + a_4 \lambda_2 + a_1 \lambda_3) = 0, \tag{48d}$$

$$x_t - x_{t-\Delta t} - \Delta x_t = 0 \tag{48e}$$

$$\dot{x}_t - (1 - a_5)\dot{x}_{t-\Delta t} + a_6 \ddot{x}_{t-\Delta t} - a_4 \Delta x_t = 0 \tag{48f}$$

$$\ddot{x}_t - (1 - a_3)\ddot{x}_{t-\Delta t} + a_2 \dot{x}_{t-\Delta t} - a_1 \Delta x_t = 0. \tag{48g}$$

The instantaneous optimal control u_t can be obtained by substituting (48a)–(48c) into (48d) and is given by

$$u_t = -R^{-1} B' L^{-1}(Q_1 x_t + a_4 Q_2 \dot{x}_t) \tag{49}$$

Observe that the control input given in (49) is of the form described in (44). Substituting (49) into (11) gives the feedback-controlled equations of motion,

$$M\ddot{x}(t) + (C + a_4 B R^{-1} B' L^{-1} Q_2)\dot{x}(t) + (K + B R^{-1} B' L^{-1} Q_1)x(t) = F(t). \tag{50}$$

From (50) one can clearly see that the effect of the control input is equivalent to adjusting the damping and stiffness matrices of the structure. Moreover, (50) can be directly solved by the Newmark integration scheme as described in (42) and (43).

REMARK 1   One may apply the control input (49) into (42f) or directly substitute into (11) to form the feedback-controlled equations of motion (50). Either case will yield the same numerical solution, if the time step and the Newmark parameters for both cases are chosen to be the same. However, in order to facilitate the stability study for the feedback-controlled system, Equation (50) is considered instead. In fact, (50) is numerically much easier to solve.   ■

3.4.2. Selection of Weighting Matrices

The proceeding derivations show that the weighting matrices Q_1, Q_2 and R are directly related to the control input. It has been pointed out by Yang *et al.*[142,143] that the instantaneous optimal control algorithm do not always guarantee the stability of the controlled structure unless the weighting matrices are chosen properly. This can also be seen from (50), in which the effect of control input may result in nonsymmetrical damping and stiffness matrices, hence it's difficult to conclude the stability of the controlled structure. A systematic way of selecting the weighting matrices through the Lyapunov method has been proposed by Yang et al.[142,143] However, the process involves solving an algebraic Riccati equation which requires substantial computational effort for a structure with high degree of freedom.

Since the external forcing function $F(t)$ is not a persistent excitation force, it is not difficult to derive a sufficient condition for the stability of (50) by properly selecting the weighting matrices. In fact, if $R > 0$, $Q_1 \geq 0$ and $Q_2 \geq 0$ are chosen such that

$$C_1' + C_1 > 0; \; K_1' = K_1 > 0, \tag{51a,b}$$

where

$$C_1 = C + a_4 B R^{-1} B' L^{-1} Q_2; \; K_1 = K + B R^{-1} B' L^{-1} Q_1. \tag{52a,b}$$

then the stability of (50) is assured while the performance index (45) is minimized.

REMARK 2   We can prove the sufficiency of condition (51) by utilizing Lyapunov stability theory, for which $V(x, \dot{x}) = x' K_1 x + \dot{x}' M \dot{x}$ is chosen as the Lyapunov function for (50). Alternatively, one can also use the properties of second-order models presented by Gardnier[157] to show the sufficiency of (51).   ■

If the weighting matrices are assumed as

$$Q_1 = \alpha_1 L; \; Q_2 = \alpha_2 L. \tag{53a,b}$$

where $\alpha_1 > 0$ and $\alpha_2 > 0$ are the weighting factors. Substituting (53) into (49) and (50), the control input and equations of motion can be written as

$$u_t = -R^{-1} B'(\alpha_1 x_t + a_4 \alpha_2 \dot{x}_t). \tag{54}$$

$$M\ddot{x}(t) + (C + a_4 \alpha_2 B R^{-1} B') \dot{x}(t) + (K + \alpha_1 B R^{-1} B') x(t) = F(t). \tag{55}$$

It can be readily seen from (55) that by assuming the weighting matrices in the forms of (53), the additional damping and stiffness matrices induced by the control input are positive semi-definite and the conditions in (51) are satisfied. As a result, the damping and stiffness of the controlled structure are increased. Of course, there are many other choices for weighting matrices.[158]

It is interesting to note that, under the condition of collocated sensors and actuators, the instantaneous optimal control derived above can be reduced to a direct output feedback. The feedback-controlled system described in (55) may be considered as a case where the control input is a direct output feedback of the form

$$u(t) = G_1 y_1(t) + G_2 y_2(t), \tag{56}$$

with

$$y_1(t) = B'x(t); \ y_2(t) = B'\dot{x}(t), \tag{57a,b}$$

where $y_1(t), y_2(t) \in \Re^m$ denote the measured displacement and velocity outputs, respectively, which are collocated with actuators. Also, the control gain matrices are corresponding to

$$G_1 = \alpha_1 R^{-1} ; \ G_2 = a_4 \alpha_2 R^{-1} \tag{58a,b}$$

Hence, the result presented here can also be considered as an instantaneous optimal direct output feedback control.

3.4.3. Numerical Results

To illustrate the current developments on the instantaneous optimal control algorithm together with the selection of the weighting matrices, the following two numerical results show the dynamic and control analyses of buildings subjected to seismic and wind loads, respectively. The time increment Δt used in the Newmark integration scheme is assumed to be 0.01 seconds. The Newmark parameters γ and β are set to be 0.5 and 0.25, respectively.

3.4.3.1. Control of a ten-story three-bay building frame

The ten-story three-bay building frame used in Section 3.3.5 is chosen for the purpose of an example study. Six different control designs involving base isolators[159] and active tendons (see Figure 20a-f) are studied: case (a) the reference frame; case (b) active control using 4 active tendons (Figure 20b); case (c) passive control using 4 base isolators; cases (d), (e), (f)-hybrid controls using 4, 1, and 5 active tendons on the base isolated frame, respectively (Figure 20d-f). The design loads and other information are as assumed in Section 3.3.5. The stiffness coefficient k_b for the elastometric base isolators used in cases (c)–(f) is assumed to be 2.74 MN/m. The damping coefficient ξ_b for the isolator is assumed to be 10%. Based on finite element modeling, the first five natural frequencies of the reference frames are

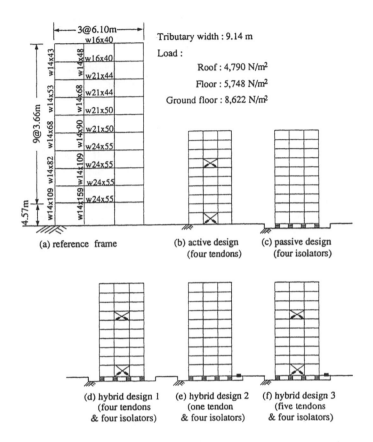

Figure 20. A 10-story 3-bay steel structural frame with various control designs.

calculated to be 0.403, 1.101, 1.866, 2.705 and 3.631 Hz for case (a) and 0.740, 2.368, 4.476, 4.521, and 6.139 Hz for case (c), respectively. In the following analyses, the weighting matrices are chosen to be

$$Q_1 = \alpha_1 L \; ; \; Q_2 = \alpha_2 L \; ; \; R = B'LB \qquad (59a,b,c)$$

Five parameters are chosen as criteria to evaluate and rate the effectiveness of the various control designs:

(i)   Maximum lateral displacement X_{max};
(ii) Maximum base displacement $(X_b)_{max}$;
(iii) Maximum inter-story drift ratio, defined as, $\frac{(X_i - X_{i-1})_{max}}{h_i}$, where h_i represents the height between the ith and $i - 1^{th}$ stories;

(iv) Maximum absolute lateral acceleration, $(\ddot{X} + \ddot{X}_g)_{\max}$;

(v) Average control force J_u, defined as, $\left(\frac{1}{t_f} \int_0^{t_f} u'u\,dt\right)^{1/2}$.

Among the five parameters, (i)–(iii) are directly related to structural safety, (iv) serves as an indicator of human reactive feeling, whereas the last one represents the amount of control force needed.

The shear beam and finite element models

The shear beam model has been commonly used in the control design and analysis of tall building frames due to its attractiveness in simplicity and practicality. However, for multiple bay structural frames with relatively flexible floor beams, more general and sophisticated structural modeling methods, such as the finite element method, may be desirable. Not only can the structural dynamic behavior be more completely modeled but also more detailed knowledge of displacements, moments, axial forces, and shear forces can be obtained. In this example, all cases are modeled using a two-node, six-degree-of-freedom planar beam-column element. Since the reference frame consists of quite a few members (30 beam and 40 column members), one element is used to model each of the 70 members. The reference frame is also studied using 140 elements, i.e., two elements for each beam or column member. However, it is found that the simpler 70 element model could generate dynamic behavior quite similar to that of the 140 element model. Thus the 70 element model is used in this example.

Figure 21 shows the maximum floor displacements and accelerations of the ten-story three-bay steel frame [case (a)] using shear beam and finite element models. It is seen that the maximum distributed floor displacements calculated using shear beam model are about one third of those calculated using finite element model. On the other hand, the maximum floor accelerations calculated using shear beam model are higher than those using the finite element model except between the 6th and the 8th floor levels. It is seen that the more flexible modeling of the frame using more degrees of freedom results in accelerations more smoothly distributed along the height as compared to the shear-beam model.

The effect of α_1 and α_2 on control efficiency

As seen from (54), the weighting factors α_1 and α_2 are directly linked to the displacement and velocity feedback, respectively. By using different values for and α_1 and α_2, the structural response can be reduced to various degrees. As a result, before any logical choices can be made between different control designs, an overall observation of the effects of weighting factors α_1 and α_2 on control efficiency must be made.

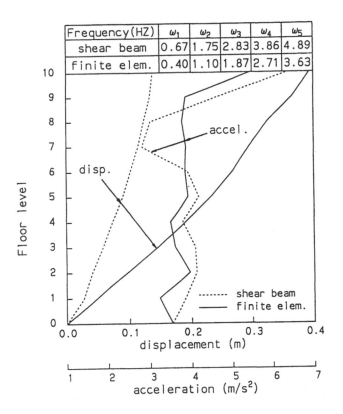

Frequency (HZ)	ω_1	ω_2	ω_3	ω_4	ω_5
shear beam	0.67	1.75	2.83	3.86	4.89
finite elem.	0.40	1.10	1.87	2.71	3.63

Figure 21. Maximum floor displacements and accelerations of the 10-story 3-bay steel frame calculated using shear beam and finite element models.

Figure 22 shows the relations between the maximum top floor displacement (X_{10}) and the average control force J_u for design case (b) for five combinations of α_1 and α_2 values: $\alpha_1 = 0$; $\alpha_1 = 10^2 \times \alpha_2$; $\alpha_1 = 10^3 \times \alpha_2$; $\alpha_1 = 10^4 \times \alpha_2$; and $\alpha_2 = 0$. The zero values for α_1 and α_2 correspond to velocity feedback and displacement feedback, respectively. It is seen that for all five sets of α_1 and α_2, the value of $(X_{10})_{max}$ decreases as J_u increases. All five curves eventually meet at the point with $((X_{10})_{max} = 0.27$ m and $J_u = 0.76$ MN, which corresponds to the result of the frame with the 1st floor rigidly bound to the ground floor, while the 7th floor is rigidly bound to the 6th. If the same amount of reduction in $((X_{10})_{max}$ is to be achieved, say, from 0.38 to 0.3 m, then the average control force required is the smallest for velocity feedback $(J_u = 0.26$ MN) and the largest for displacement feedback $(J_u = 0.68$ MN).

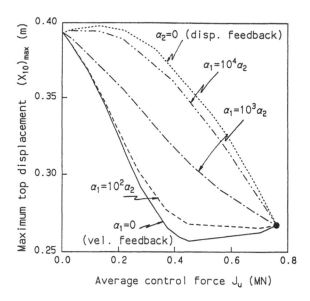

Figure 22. Control efficiency of design case (b) (4 active tendons) with 5 different a_1 and a_2 combinations.

This observation is also true for other levels of reduction in $(X_{10})_{\max}$. The two curves for $\alpha_1 = 0$ and $\alpha_2 = 0$ are the envelope of the other three curves. This seems to suggest that velocity feedback ($\alpha_1 = 0$) is more efficient than the other combinations of α_1 and α_2 values as it requires the least amount of average control force for the same amount of reduction in displacement everywhere in the figure.

The time histories of the top floor displacement X_{10} and the control force in the active tendons between the ground and first floor using displacement and velocity feedback with $(X_{10})_{\max} = 0.3$ m are shown in Figure 23. It is seen that although the α_1 and α_2 values for the displacement feedback ($\alpha_1 = 0.2$) and the velocity feedback ($\alpha_2 = 3 \times 10^{-5}$) are deliberately chosen so that they give the same $(X_{10})_{\max}$ value, the displacement peaks using the displacement feedback are generally higher than those using the velocity feedback. The control forces required using the displacement feedback are also higher than those using the velocity feedback. It appears that the velocity feedback is more advantageous than the displacement feedback.

Figure 24 shows the control efficiency in terms of the relations between the maximum top floor displacement $(X_{10})_{\max}$ and base floor displacement $(X_b)_{\max}$ and the average control force J_u for the hybrid design cases (d), (e),

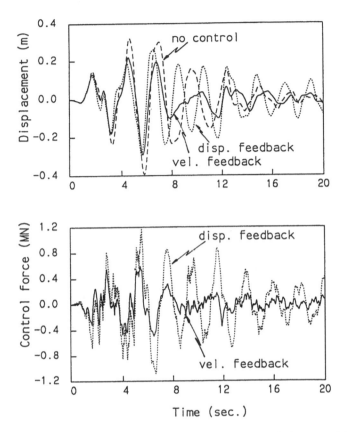

Figure 23. Time histories of the top floor displacement and the control force of the active tendons between ground and first floor using displacement and velocity feedback in Figure 22 with $(x_{10})_{max} = 0.3$ m.

and (f) using displacement and velocity feedback, respectively. Again in this case, the velocity feedback seems to be more efficient than the displacement feedback. It is also noted that the control forces required for cases (c), (d), (e), and (f) all fall within the operational range of some practical hydraulic actuators, such as the MTS model series 244 and 247 which have a force rating up to 1 MN and 1.1 MN, respectively.[160]

Comparative studies of the six design cases

An effort is made here to study the comparative performance of the six design cases. Such a study is very complex because it involves too many

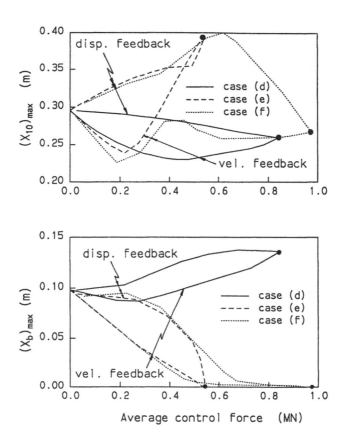

Figure 24. Control efficiency of hybrid design cases (d), (e), and (f) using displacement and velocity feedback.

design parameters. In order to avoid a lengthy and complex parametric study and yet to aim at drawing some meaningful and useful conclusions and recommendations, a comparative study with a limited number of carefully selected parameter values is conducted. In such a study, some parameters, such as the stiffness and damping coefficients of the base isolators are pre-determined, while other parameters, such as the weighting parameters, are assumed to have specific values. Thus it is assumed that, in design cases (b), (d), (e), and (f), only velocity feedback is used ($\alpha_1 = 0$) and α_2 is assumed to be 3×10^{-5}, 3.6×10^{-5}, 3.5×10^{-6} and 5×10^{-6}, respectively. With these values of α_2, the average control force required for the four cases is all equal to 0.28 MN.

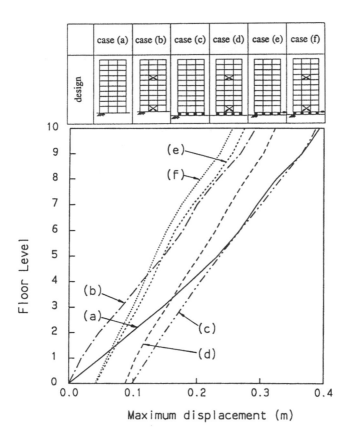

Figure 25. Maximum floor displacements for the six design cases.

Figure 25 shows the maximum lateral displacements of the frame relative
to the ground for these six design cases. The distribution of the maximum
displacements for case (b) are about 75% of those of case (a). For the four
cases with base isolators, case (c) (with isolators only) shows the largest
base displacement. With 4, 1, and 5 active tendons on the isolated frame,
respectively, cases (d), (e), and (f) demonstrate various degrees of reduction
in the base displacement. Among them, cases (e) and (f) have almost the
same magnitude and trend of reduction. The maximum drift ratios for these
six cases are plotted in Figure 26. As compared to those of case (a), all other
five cases show significant reductions to various extents (between 40–60%).
Among the five cases, case (d) seems to have the most uniform reductions. It
is noted that the drift ratio between the ground and 1st floor of case (e) is the

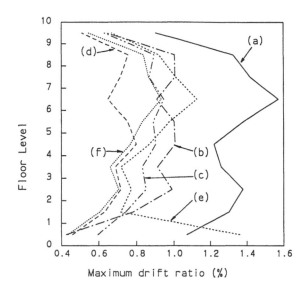

Figure 26. Maximum inter-story drift ratios.

largest among the six cases considered. This is attributed to the fact that there is only one control force acting at the ground floor in the direction opposite to the motion of the isolated frame, thereby increasing the drift ratio between ground and 1st floor. Figure 27 shows the maximum absolute accelerations for the six cases. It is seen that as compared to case (a), all other five cases exhibit reduction of the maximum absolute accelerations to various degrees. Among them, case (d) has the largest reduction, followed by case (c). The absolute accelerations for cases (e) and (f) are higher than those of (c) and (d). This seems to be the result of applying an active tendon at the ground floor level which reduces the ground floor displacement but increases the absolute accelerations of the frame.

While the shear-beam model remains perhaps the most popular structural modeling method for dynamics and control studies of tall building frames for its simplicity and often acceptable accuracy, the computationally more expensive finite element model could complement the shear-beam model with some additional advantages. As mentioned before, it gives a more complete and accurate dynamic knowledge of the frame. Figures 28–30 show the distributions of maximum axial forces, shear forces, and moments along the outermost columns of the frame for the six designs, respectively. As compared to those of case (a), all the other five cases demonstrate reductions in axial

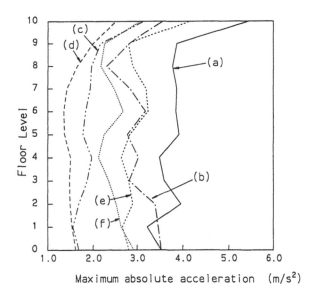

Figure 27. Maximum absolute floor accelerations.

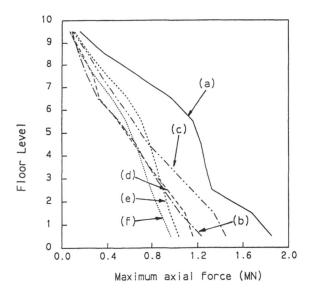

Figure 28. Maximum axial forces in the outer columns of the steel frame.

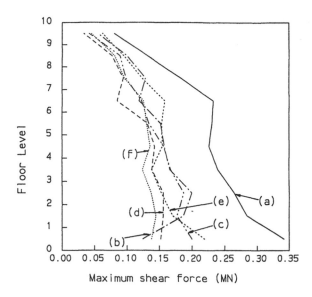

Figure 29. Maximum shear forces in the outer columns of the steel frame.

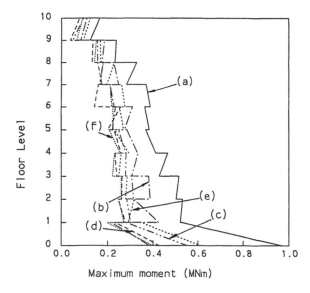

Figure 30. Maximum moments in the outer columns of the steel frame.

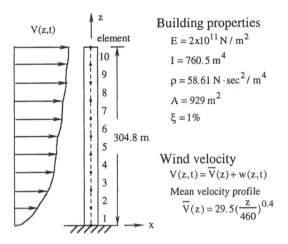

Figure 31. Building and wind velocity models.

force, shear force, and moment distributions to various degrees. It is difficult to conclude as to which case performs the best since these distribution curves are intertwined. However, the general distribution trends for all five cases seem to be similar. It also appears that cases (d) and (f) are slightly better than the others as they have the lowest and most uniform maximum axial forces, shear forces, and moments.

3.4.3.2. Control of a tall building

In this example, dynamic and control analyses of a tall building subjected to wind loads are performed. The building is simplified as a cantilever with square cross-section (see Figure 31). The structural properties of the building are assumed as modulus of elasticity $E = 2 \times 10^{11}$ N/m^2, moment of inertia $I = 760.5$ m^4, density $\rho = 58.61$ N-sec^2/m^4, height $L = 304.8$ m, cross section area $A = 929.0$ m^2.[161] The damping coefficients are assumed as 1% for all modes. Ten two-node, six-degree-of-freedom beam elements are used to model the building. The first natural frequency of the building is calculated as 2 rad/sec. The wind velocity is assumed to be random in both one dimensional space (in z direction only) and time domains. A power law representation, as suggested by Davenport[162], for the mean velocity profiles $\overline{V}z$ of the form

$$\overline{V}(z) = \overline{V}_G \left(\frac{z}{z_G} \right)^{\alpha} \tag{60}$$

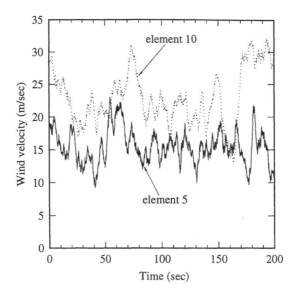

Figure 32. Wind velocity histories on elements 5 and 10.

is used. In this study, the parameters are assumed as, the exponent $\alpha = 0.4$, the reference height $Z_G = 460$ m, and the reference wind velocity $\overline{V}_G = 29.5$ m/sec. The turbulence component of the velocity w is expressed in the Simiu spectrum[163] as

$$S(z, n) = \frac{V_s^2}{n} \frac{200\hat{f}}{(1 + 50\hat{f})^{5/3}} \tag{61}$$

where $S(z, n)$ is the power spectral density of the wind turbulence component at location z and frequency n, V_s is the shear velocity and assumed as 0.91 m/sec, $\hat{f}$ denotes the reduced frequency defined as

$$\hat{f} = \frac{nz}{\overline{V}(z)}. \tag{62}$$

The time domain of the turbulence component w is then simulated using the spectral representation method developed by Shinozuka, *et al.*[164]

Figure 32 shows the 200 seconds generated wind velocity histories on elements 5 and 10. The first 100 seconds of the displacement and acceleration histories at the top of the building subjected to this set of simulated wind loads are shown in Figure 33. It is seen that the displacement response is basically contributed from the first mode of the building that vibrates with respect to

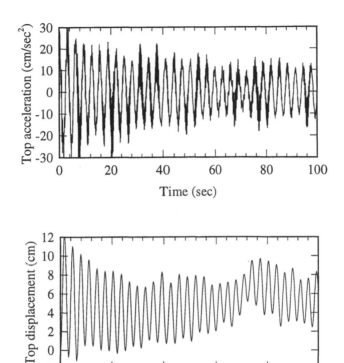

Figure 33. Top displacement and acceleration histories of the building subjected to wind loads.

a neutral value of 5 cm. On the other hand, the acceleration response comes not only from the first mode but also from the combination of other higher modes.

In the following analyses, the time average quantities defined in (41), such as the average displacement J_x, the average acceleration $J_{\ddot{x}}$, and the average control force J_u are used for discussion. Two sets of weighting matrices are used in the analyses. The first set is as given in (59) and the second set is chosen to be

$$Q_1 = \alpha_1 K \; ; \; Q_2 = \alpha_2 M \; ; \; R = B'K^{-1}B. \qquad (63a,b,c)$$

The second set of weighting matrices is selected so that the first two terms on the right hand side of the performance index (45) represent the

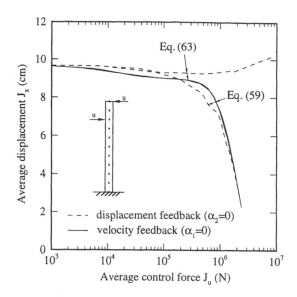

Figure 34. Average displacement of the building J_x versus average control force J_u with weighting matrices assumed in the forms of Equation (59) and Equation (63), respectively.

structural strain and kinetic energies, respectively, and the third term denotes the quasistatic work done by the control force.

Figure 34 shows the average displacement of the building J_x versus average control force J_u for the case of one control force acting on top of the building and at top of the eighth element with the same magnitude but opposite in direction. This control force represents a cross-diagonal tendon type of actuator. Two different conditions are considered, one corresponds to velocity feedback ($\alpha_1 = 0$), the other corresponds to displacement feedback ($\alpha_2 = 0$). It is seen that among these four curves, three curves follow a similar trend, i.e., J_x reduces as J_u increases. Intuitively, this also says that the more the control effort (J_u) put in, the better the suppression of the response (J_x). However, the average displacement J_x resulted from using displacement feedback with the weighting matrices (63) follows a different trend as the average control force J_u increases. In fact, the numerical calculation reveals that J_x in this case actually becomes unbounded as J_u increases. This unstable result confirms the argument stated previously that, in the problem of instantaneous optimal control, it is not sufficient from the viewpoint of closed-loop stability to select the weighting matrices to be just positive definite as in the case of (63).

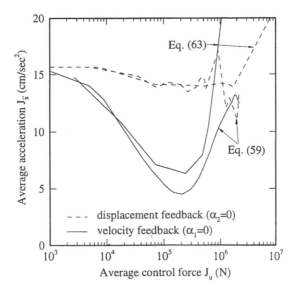

Figure 35. Average acceleration of the building $J_{\ddot{x}}$ versus average control force J_u for the case of Figure 34.

The results of the average acceleration $J_{\ddot{x}}$ versus the average control force J_u for the case of Figure 34 is plotted in Figure 35. It is seen that the results using velocity feedback ($\alpha_1 = 0$) follow a similar trend for both selections of weighting matrices, i.e., $J_{\ddot{x}}$ reduces first when J_u increases and reaches its minimum when J_u equals approximately 2×10^5 N. On the other hand, in the case of displacement feedback with the weighting matrices chosen in the form of (59), the average acceleration $J_{\ddot{x}}$ shows small fluctuation for J_u up to 5×10^5 N, whereas with the weighting matrices chosen according to (63), the average acceleration $J_{\ddot{x}}$ increases sharply as J_u is larger than 10^6 N.

In addition to the case studied in Figures 34 and 35, three more cases showing different actuator designs (see Figure 36) are analyzed. In case 2, five cross diagonal actuators are placed in the building; case 3 represents the situation where only one control force is placed on top of the building; while case 4 corresponds to putting one control force on top of elements 2, 4, 6, 8, and 10, respectively. The results of J_x versus J_u and $J_{\ddot{x}}$ versus J_u for these four cases are plotted in Figures 36 and 37, respectively. It is seen from Figure 36 that the average displacement J_x for all four cases using either displacement or velocity feedback reduces as J_u increases. Among the four cases, it seems that case 4 is the most efficient one since for the same amount of average control force, it yields the lowest average displacement.

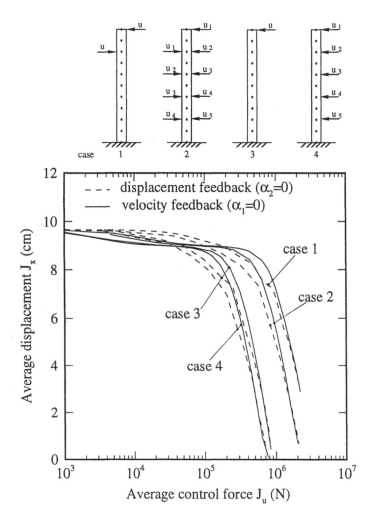

Figure 36. Average displacement of the building J_x versus average control force J_u for four actuator design cases.

As for the curves of $J_{\ddot{x}}$ versus J_u for the four cases shown in Figure 37, the curves corresponding to velocity feedback first reduce to certain values and then increase thereafter. On the other hand, the results of displacement feedback do not show any significant reduction. The closed-loop stability of the four cases using either velocity or displacement feedback is numerically verified in these two figures.

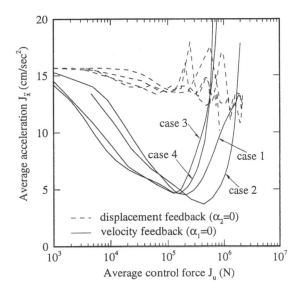

Figure 37. Average acceleration of the building $J_{\ddot{u}}$ versus average control force J_u for four actuator design cases.

3.4.3.3. Further comparisons on the control efficiency

As demonstrated in Sections 3.5 and 4.3.1, to suppress vibration of a ten-story three-bay building frame, the control forces can be designed using either the instantaneous optimal control algorithm which basically corresponds to the tendon-type of controller or the complete feedback control algorithm which is used with a TMD system. To make a comparison on the control efficiency between these two algorithms, the same building frame subjected to the NS component of the ground acceleration of the 1940 El Centro earthquake is used again to study the following three cases: case (a) instantaneous optimal control using 4 tendons; case (b) instantaneous optimal control using 4 tendons together with a TMD system; and case (c) complete feedback control using an ATMD system. The mass of the TMD system used in cases (b) and (c) is assumed to be 1% of the total mass of the building frame.

Figure 38 shows the maximum top displacement versus the average control force J_u for these three design cases. Both displacement feedback and velocity feedback are used for design cases (a) and (b). It can be seen that, between these two cases, the control forces required using velocity feedback are smaller than those using displacement feedback. Among these

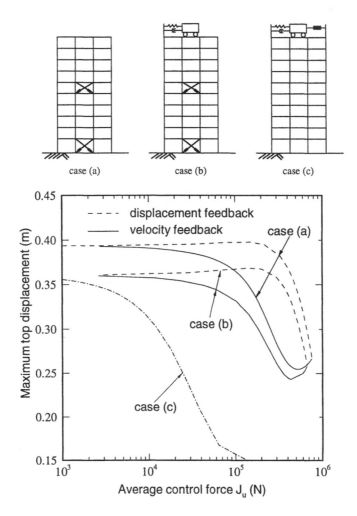

Figure 38. Maximum top displacement versus average control force for design cases (a), (b), and (c).

three design cases, it seems that case (c) is more efficient than the other two cases. For example, to reduce the maximum top displacement of the building frame to 0.3 m, it requires average control force of 2.5×10^5, 1.9×10^5, and 1.03×10^4 for cases (a), (b), and (c), respectively.

The relations between the maximum top absolute acceleration and the average control force J_u for the three design cases are plotted in Figure 39.

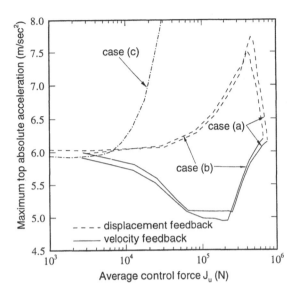

Figure 39. Maximum top absolute acceleration versus average control force for the three design cases.

It shows that the top absolute accelerations decrease initially from 6 to 5 m/sec^2 for design cases (a) and (b) using velocity feedback as J_u increases. However, as J_u increases beyond 2.5×10^5 N, an increasing trend is observed for both cases (a) and (b). For design case (c), the top absolute acceleration increases monotonically as J_u increases. This phenomenon demonstrates that as far as the suppression of absolute acceleration is concerned, the complete feedback algorithm not only can not reduce the acceleration but would increase it instead.

Figures 40 and 41 show the maximum floor displacement profiles and the maximum absolute floor acceleration profiles for the three design cases, respectively. The parameters for these three design cases are selected so that the maximum top displacements are all equal to 0.3 m. The velocity feedback control is used for design cases (a) and (b). It is seen that the results of design case (c) are higher than those of cases (a) and (b) at most of the floor levels. Also, the displacement profile and the acceleration profile for design cases (a) and (b) are almost the same. As far as the response of the TMD system is concerned, these two figures show that the TMD stroke and acceleration are both higher for design case (c) than for case (b).

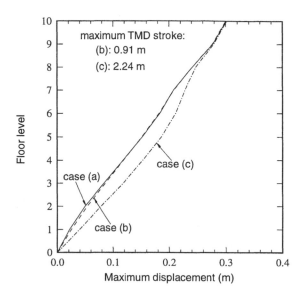

Figure 40. Maximum floor displacements for the three design cases.

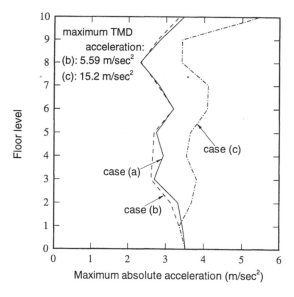

Figure 41. Maximum absolute accelerations for the three design cases.

In summary, although the complete feedback control using an ATMD system is more efficient in suppressing the structural displacement, it results in significant increases of both structural acceleration and the TMD responses. Also, the combined usage of a TMD system with active tendons based on the instantaneous optimal control algorithm can produce satisfactory suppressing results and requires less amount of control energy as compared to using active tendons alone.

3.5. CONCLUDING REMARKS

In this chapter, two techniques which can be used for the control of civil engineering structures were presented.

First, a closed-loop complete feedback control algorithm was proposed for the control of a structure modeled as a SDOF system using an ATMD. The SDOF system was assumed to be under stationary Gaussian white noise ground excitation. The control force was calculated from the acceleration, velocity and displacement feedback of the structure and the active mass damper. The properties of the tuned mass damper and the gain coefficients of the actuator were derived by minimizing the displacement variance of the SDOF system. The stability of the proposed algorithm was also discussed using the Routh-Hurwitz criterion. Numerical simulations were performed to evaluate the performance of the ATMD design. The results demonstrated that by placing an actuator, which generates the control force using velocity feedback, between the structure and the TMD with optimal parameters not only fail to reduce the response of the structure, but increase the response instead. The structural responses could be reduced either by using velocity feedback on an ATMD with non-optimal parameters or by using complete feedback on an ATMD with optimal parameters. The reduction of the structural response, however, was accompanied by the increase of the TMD response. Further comparison study demonstrated that the ATMD with optimal parameters using complete feedback was more efficient than the ATMD with non-optimal parameters using velocity feedback.

Also, a control algorithm based on the instantaneous optimal closed-loop control technique was presented for the control of a ten-story three-bay building frame subjected to earthquake excitations and a tall building subjected to random wind loads. The control design procedure involved only straightforward algebraic manipulations, which is simpler than the Riccati equation approach, especially when dealing with a structure of high degrees of freedom.

In the problem of instantaneous optimal control, the weighting matrices have to be selected cautiously. In fact, it was shown through numerical simulations that although the weighting matrices were chosen to be positive definite, the resultant feedback-controlled equations of motion was unstable.

An attempt was made to study the control performances of many different control designs which are considered practical. While the current numerical results do not clearly conclude as to which control design or what values of parameters or coefficients are the best for design purposes, they serve to demonstrate the current technique of instantaneous optimal closed-loop control based on a finite element formulation. It is felt that the current study may, nonetheless, provide some insight into the physical behavior as well as the characteristics of the control devices studied.

References

1. Liu, S.C., H.J. Lagorio, and K.P. Chong, 1991, *Earthquake Spectra*, **7**, 543–550.
2. Newsletter, 1995, *Int. Assoc. for Struct. Cont.*, **1**.
3. Zuk, W., 1968, *Civil Engineering* (ASCE), 62–64.
4. Nordell, W.J., 1969, *Tech. Report* P-611, Naval Civil Eng. Lab., CA.
5. Murata, M. and N. Ito, 1971, *Proc. 3rd Int. Conf. Wind Effects on Buildings and Struct.*, 1057–1066.
6. Yao, J.T.P., 1969, *Report on NSF-UCEER Earthquake Engrg. Research Conf.*, Univ. of California, Berkeley, CA.
7. Yeh, H.Y. and J.T.P Yao, 1969, *ASME Preprint No.* 69-VIBR-20.
8. Yao, J.T.P, 1972, *J. Struct. Division (ASCE)*, **98**, 1567–1574.
9. Yao, J.T.P and J.P. Tang, 1973, *Tech. Report No.* CE-STR-73–1, Purdue Univ.
10. Yao, J.T.P and S. Sae-Ung, 1978, *J. Engrg. Mech. (ASCE)*, **104**, 335–350.
11. Martin, C.R. and T.T. Soong, 1976, *J. Engrg. Mech. (ASCE)*, **102**, 613–632.
12. Yang, J.N. and J.T.P. Yao, 1974, *Tech. Report No.* CE-STR-74–2, Purdue Univ.
13. Yang, J.N., 1975, *J. Engrg. Mech. (ASCE)*, **101**, 818–838.
14. Roorda, J., 1975, *J. Struct. Division (ASCE)*, **101**, 505–521.
15. Abdel-Rohman, M. and H.H.E. Leipholz, 1979, *J. Engrg. Mech. (ASCE)*, **105**, 1007–1923.
16. Yang, J.N. and B. Samali, 1983, *J. Struct. Engrg. (ASCE)*, **109**, 50–68.
17. Yang, J.N. and F. Giannopoulos, 1979, *J. Engrg. Mech. (ASCE)*, **105**, 677–694.
18. Roorda, J., 1980, *Structural Control* (North-Holland Pub. Co.), pp. 629–662.
19. Chang, J.C.H. and T.T. Soong, 1980, *J. Engrg. Mech. (ASCE)*, **106**, 1091–1098.
20. Klein, R.E. and H. Salhi, 1980, *Structural Control* (North-Holland Pub. Co.), pp. 415–430.
21. Chang, J.C.H. and T.T. Soong, 1980, *Structural Control* (North-Holland Pub. Co.), pp. 199–210.
22. T.T. Soong and G.T. Skinner, 1981, *J. Engrg. Mech. (ASCE)*, **107**, 1057–1068.
23. Soong, T.T. and M.I.J. Chang, 1980, *Structural Control* (North-Holland Pub. Co.), pp. 723–738.
24. Vilnay, O., 1981, *J. Engrg. Mech. (ASCE)*, **107**, 907–916.
25. Yang, J.N. and M.J. Lin, 1981, *Proc. ASCE EMD/STD Symp. Prob. Meth. in Struct. Engrg.*, pp. 102–120.
26. Udwadia, F.E. and S. Tabaie, 1981, *J. Engrg. Mech. (ASCE)*, **107**, 997–1009.
27. Abdel-Rohman, M. and H.H.E. Leipholz, 1981, *J. Struct. Division (ASCE)*, **107**, 1313–1325.
28. Abdel-Rohman, M., V. Quintana, and H.H.E. Leipholz, 1980, *J. Engrg. Mech. (ASCE)*, **106**, 57–73.
29. Robinson, A.C., 1971, *Automatica*, **7**, 371–388.

30. Yang, J.N., A. Akbarpour, and P. Ghaemmaghami, 1987, *J. Engrg. Mech. (ASCE)*, **113**, 1369–1386.
31. Yang, J.N., A. Akbarpour, and P. Ghaemmaghami, 1980, *Structural Control* (North-Holland Pub. Co.), pp. 748–761.
32. Chung, L.L., R.C. Lin, T.T. Soong, and A.M. Reinhorn, 1989, *J. Engrg. Mech. (ASCE)*, **115**, 1609–1627.
33. Lin, R.C., T.T. Soong, and A.M. Reinhorn, 1987, *Proc. US-Japan Joint Seminar on Stoch. Struct. Mech.*, Florida Atlantic Univ., FL.
34. Porter, B. and T.R. Crossley, 1972, *Modal Control: Theory and Applications* (Taylor and Francis, London).
35. Abdel-Rohman, M. and H.H.E. Leipholz, 1978, *J. Engrg. Mech. (ASCE)*, **104**, 1157–1175.
36. Yang, J.N. and F. Giannopoulos, 1978, *J. Engrg. Mech. (ASCE)*, **104**, 551–568.
37. Yang, J.N. and M.J. Lin, 1983, *J. Engrg. Mech. (ASCE)*, **109**, 1375–1389.
38. Yang, J.N., 1982, *J. Engrg. Mech. (ASCE)*, **108**, 833–849.
39. Yang, J.N. and F. Giannopoulos, 1979, *J. Engrg. Mech. (ASCE)*, **105**, 795–810.
40. Yang, J.N. and M.J. Lin, 1982, *J. Engrg. Mech. (ASCE)*, **108**, 1167–1185.
41. Meirovitch, L. and H. Oz, 1980, *J. Guidance, Cont. and Dyn. (AIAA)*, **3**, 140–150.
42. Meirovitch, L. and H. Baruh, 1982, *J. Guidance, Cont. and Dyn. (AIAA)*, **5**, 60–66.
43. Meirovitch, L. and H. Baruh, 1983, *J. Guidance, Cont. and Dyn. (AIAA)*, **6**, 20–25.
44. Meirovitch, L. and L.M. Silverberg, 1983, *J. Engrg. Mech. (ASCE)*, **109**, 604–618.
45. Meirovitch, L. and L.M. Silverberg, 1983, *Opt. Cont. Appli. and Meth.*, **4**, 365–386.
46. Meirovitch, L., 1983, *J. Engrg. Mech. (ASCE)*, **109**, 604–618.
47. Meirovitch, L. and D. Ghosh, 1987, *J. Engrg. Mech. (ASCE)*, **113**, 720–736.
48. Meirovitch, L., 1987, *J. Opt. Theory and Appli.*, **54**, 1–22.
49. Meirovitch, L. and H. Oz, 1980, *Structural Control* (North-Holland Pub. Co.), pp. 505–521.
50. Berry, D.T., T.Y. Yang, and R.E. Skelton, 1985, *J. Guidance, Cont. and Dyn. (AIAA)*, **5**, 612–619.
51. Lamberson, S. and T.Y. Yang, 1985, *Computers and Structures*, **20**, 583–592.
52. Lamberson, S. and T.Y. Yang, 1986, *J. Guidance, Cont. and Dyn. (AIAA)*, **9**, 476–484.
53. Hu, A., R.E. Skelton, and T.Y. Yang, 1987, *J. Sound and Vib.*, **117**, 475–496.
54. Prucz, Z. and T.T. Soong, 1983, *Recent Adv. in Engrg. Mech. and Their Impact on Civil Engrg. Prac.*, edited by W.F. Chen and A.D.M. Lewis, pp. 903–906.
55. Reinhorn, A.M., T.T. Soong, and G.D. Manolis, 1986, *Proc. ASME 5th Int. OMAE Conf.*, pp. 39–44.
56. Masri, S.F., G.A. Bekey, and T.K. Caughey, 1981, *J. Appl. Mech. (ASME)*, **48**, 619–626.
57. Masri, S.F., G.A. Bekey, and T.K. Caughey, 1982, *J. Appl. Mech. (ASME)*, **49**, 877–884.
58. Udwadia, F.E. and S. Tabaie, 1981, *J. Engrg. Mech. (ASCE)*, **107**, 1011–1028.
59. Prucz, Z., T.T. Soong, and A.M. Reinhorn, 1985, *J. Dyn. Sys. Meas. and Cont. (ASME)*, **107**, 123–131.
60. Masri, S.F., G.A. Bekey, and F.E. Udwadia, 1980, *Structural Control* (North-Holland Pub. Co.), pp. 471–492.
61. Reinhorn, A.M., T.T. Soong, and C.Y. Wen, 1987, *Proc. ASME PVP Conf.*), San Diego, CA.
62. Lee, S.K. and F. Kozin, 1986, *Dyn. Resp. of Struct.*, edited by G.C. Hart and R.B. Nelson, pp. 788–794.
63. Lee, S.K. and F. Kozin, 1987, *Structural Control* (Martinus Nijhoff Pub. Co.), pp. 387–407.
64. Samali, B., J.N. Yang, and C.T. Yeh, 1985, *J. Engrg. Mech. (ASCE)*, **111**, 777–796.
65. Samali, B., J.N. Yang, and S.C. Liu, 1985, *J. Struct. Engrg. (ASCE)*, **111**, 2165–2180.
66. Yang, J.N., A. Akbarpour, and P. Ghaemmaghami, 1987, *Tech. Report NCEER-TR-87-0007*, Nat. Center for Earthquake Eng. Research.
67. Prucz, Z., T.T. Soong, and A.M. Reinhorn, 1984, *ASCE Annual Convention*, Atlanta, GA.
68. Ghaemmaghami, P. and J.N. Yang, 1985, *Proc. 4th Int. Conf. on Struct. Rel. and Safety*, pp. III-213–222.
69. Ghaemmaghami, P. and J.N. Yang, 1987, *Proc. US-Austria Joint Sem. on Stoch. Struct. Dyn.*, Florida Atlantic Univ., FL.

70. Reinhorn, A.M. and G.D. Manolis, 1985, *The Shock and Vib. Digest*, **17**, 35–41.
71. Soong, T.T., 1988, *Engrg. Struct.*, **10**, 74–84.
72. Yang, J.N. and T.T. Soong, 1988, *Prob. Engrg. Mech.*, **3**, 179–188.
73. Soong, T.T., 1990, *Active Struct. Cont.: Theory and Practice* (Longman, London, and Wiley)
74. McNamara, R.J., 1977, *J. Struct. Division (ASCE)*, **103**, 1785–1798.
75. Luft, R.W., 1979, *J. Struct. Division (ASCE)*, **105**, 2766–2772.
76. Warburton, G.B. and E.O. Ayorinde, 1980, *Earthquake Engrg. and Struct. Dyn.*, **8**, 197–217.
77. *Engineering News Record*, 1976, **197**, 10.
78. Isyumov, N., J. Holmes, and A.G. Davenport, 1975, *Univ. of Western Ontario Research Report BLWT-551–75*, London, Ontario, Canada.
79. *Engineering News Record*, 1979, **200**, 11.
80. Naruse, T. and Y. Hirashima, 1987, *STAHLBAU*, **7**, 193–196.
81. Nishino, H., T. Mochio, and T. Hara, 1993, *Proc. Asia-Pacific Vib. Conf.*, Kitakyushu, Japan, pp. 283–287.
82. Yoshida, O. and S. Kaneko, 1993, *Proc. Asia-Pacific Vib. Conf.*, Kitakyushu, Japan, pp. 312–317.
83. Mizota, Y., S. Kaneko, 1993, *Proc. Asia-Pacific Vib. Conf.*, Kitakyushu, Japan, pp. 318–323.
84. Samali, B., K.C.S. Kwok, and D. Tapner, *Proc. IABSE Congress*, pp. 461.
85. Hitchcock, P.A., K.C.S. Kwok, R.D. Watkins, and B. Samali, 1993, *Proc. Asia-Pacific Vib. Conf.*, Kitakyushu, Japan, pp. 799–803.
86. Izumi, M., T. Teramoto, H. Kitamura, and Y. Shirasawa, 1993, *Proc. Struct. Congress* (Irvine, CA), edited by A.H-S. Ang and R. Villaverde, pp. 107–114.
87. Fujino, Y. and M. Matsumoto, 1993, *Proc. Struct. Congress* (Irvine, CA), edited by A.H-S. Ang and R. Villaverde, pp. 544–549.
88. Keel, C.J. and P. Mahmoodi, 1986, *Proc. Building Motion in Wind*, edited by N. Isyumov and T. Tschanz, pp. 67–82.
89. Keel, C.J., 1987, *Proc. ASME Design Tech. Conf.*, edited by L. Rogers, pp. 28–1–2–823.
90. Caldwell, D.G., 1986, *Engineering Journal (AISC)*, 148–150.
91. Pall, A.S. and C. Marsh, 1982, *J. Struct. Division (ASCE)*, **108**, 1313–1323.
92. Filiatrault, A. and S. Cherry, 1988, *Earthquake Engrg. and Struct. Dyn.*, **16**, 389–416.
93. Pekau, O.A. and R. Guimond, 1991, *Earthquake Engrg. and Struct. Dyn.*, **20**, 505–521.
94. Aguirre, M. and A.R. Sanchez, 1992, *J. Struct. Engrg. (ASCE)*, **118**, 1150–1171.
95. Seismic Resp. Cont. Series, *Tech. Pamphlet 91–64E*, Kajima Corporation, Japan.
96. Seismic Resp. Cont. Series (Joint Damper System), *Tech. Pamphlet 91–62E*, Kajima Corporation, Japan.
97. Kelly, J.M., 1990, *Earthquake Spectra*, **6**, 223–244.
98. Mostaghel, N. and M. Khodaverdian, 1987, *Earthquake Engrg. and Struct. Dyn.*, **15**, 379–390.
99. Kelly, J.M. and K.E. Beuke, 1983, *Earthquake Engrg. and Struct. Dyn.*, **11**, 33–56.
100. Kelly, J.M. and S.B. Hodder, *Bulletin New Zealand Natl. Soc. Earthquake Engrg.*, **1**, 53–67.
101. Stanton, J. and C. Roeder, 1991, *Earthquake Spectra*, **7**, 301–323.
102. Kelly, J.M., G. Leitmann, and A.G. Soldatos, 1987, *J. Opt. Theory and Appli.*, **53**, 159–180.
103. Kobori, T., 1990, *Proc. U.S. Nat. Workshop on Struct. Cont. Research*, edited by G.W. Housner and S.F. Masri, pp. 1–21.
104. Koizumi, T., Y. Furuishi, and N. Tsujiuchi, 1989, *Trans. of JSME*, **55**, 1602–1608.
105. Kobori, T., 1987, *Proc. Annual Meeting Arch. Inst.* (1987), Tokyo, Japan.
106. Kobori, T., H. Kanayama, and S. Kamagata, *Proc. 9th World Conf. on Earthquake Engrg.*, Tokyo/Kyoto, Japan, pp. 465–470.

107. Kobori, T., M. Sakamoto, M. Takahashi, N. Koshika, and K. Ishii, 1990, *Proc. U.S. Nat. Workshop on Struct. Cont. Research*, edited by G.W. Housner and S.F. Masri, Univ. of Southern California, Los Angeles, CA.

108. Kobori, T., M. Sakamoto, M. Takahashi, N. Koshika, and K. Ishii, 1992, *Proc. 10th World Conf. on Earthquake Engrg.*, Madrid, Spain, pp. 1–6.

109. *Technical Pamphlet* 91–63E (1991), Kajima Corporation, Japan.

110. Torkamani, M.A. and E. Pramono, 1985, *J. Struct. Engrg. (ASCE)*, **111**, 805–825.

111. Chung, L.L., A.M. Reinhorn, and T.T. Soong, 1986, *Proc. ASCE Specialty Conf. on Struct. Dyn.*, Los Angeles, CA, pp. 795–802.

112. Chung, L.L., A.M. Reinhorn, and T.T. Soong, 1988, *J. Engrg. Mech. (ASCE)*, **114**, 241–256.

113. Lin, R.C., T.T. Soong, and A.M. Reinhorn, 1987, *Tech. Report NCEER-TR*-87–0002, Nat. Center for Earthquake Eng. Research.

114. McGreevy, S., 1987, *MS Thesis*, State Univ. of New York at Buffalo, NY.

115. Soong, T.T., A.M. Reinhorn, and J.N. Yang, 1987, *Structural Control* (Martinus Nijhoff Pub. Co.), pp. 669–693.

116. Soong, T.T., A.M. Reinhorn, Y.P. Wang, and R.C. Lin, 1991, *J. Struct. Engrg. (ASCE)*, **117**, 3516–3536.

117. Klein, R.E., C. Cusano, and J.V. Slukel, 1972, *ASME Annual Meeting*, New York.

118. Abdel-Rohman, M., 1984, *J. Struct. Engrg. (ASCE)*, **110**, 937–946.

119. Reinhorn, A.M. and G.D. Manolis, 1985, *Proc. 2nd Int. Symp. on Struct. Cont.*, Univ. of Waterloo, Ontario, Canada.

120. Reinhorn, A.M., G.D. Manolis, and C.Y. Wen, 1987, *J. Engrg. Mech. (ASCE)*, **113**, 315–333.

121. Dehghanyar, T.J., S.F. Masri, R.K. Miller, G.A. Bekey, and T.K. Caughey, 1983, *Proc. 4th VPI & SU/AIAA Symp. on Dyn. and Cont. of Large Struct.*, Blacksburg, VA.

122. Dehghanyar, T.J., S.F. Masri, R.K. Miller, G.A. Bekey, and T.K. Caughey, 1985, *Proc. NASA / JPL Workshop on Iden. and Cont. of Flexible Space Struct.*, San Diego, CA.

123. *Technical Pamphlet* 91–65E (1991), Kajima Corporation, Japan.

124. Kobori, T. *et al.*, 1990, *Proc. 4th World Cong. of Council on Tall Buildings and Urban Habitat*.

125. Kobori, T. *et al.*, 1991, *Proc. 2nd Conf. on Tall Buildings in Seismic Regions*, Los Angeles, CA, pp. 213–222.

126. Rogers, C.A., 1990, *Proc. Int. Workshop on Intel. Struct.*, Taipei, Taiwan, pp. 3–41.

127. Kitamura, H., S. Kawamura, M. Yamada, and S. Fuji, 1990, Proc. *U.S. Nat. Workshop on Struct. Cont. Research*, edited by G.W. Housner and S.F. Masri, pp. 141–150.

128. Chang, J.C.H. and T.T. Soong, 1980, *J. Engrg. Mech. (ASCE)*, **106**, 1091–1098.

129. Hrovat, D., P. Barak, and M. Rabins, 1983, *J. Engrg. Mech. (ASCE)*, **109**, 691–705.

130. Stephens, L.S., K.E. Rouch, and S.G. Tewani, 1991, *Proc. ASME Design Tech. Conf.*, edited by T.C. Huang, H.S. Tzou, P. Bainum, K.J. Saczalski, S.H. Sung, B.K. Wada, and B.P. Wang, Miami, FL, pp. 89–94.

131. Ohsaki, Y., 1990, *Proc. U.S. Nat. Workshop on Struct. Cont. Research*, edited by G.W. Housner and S.F. Masri, pp. 179–188.

132. Tanida, K., Y. Koike, M. Mutaguchi, and N. Uno, 1991–92, *Trans. of JSME*, **57**, 486–490 (in Japanese).

133. Fujino, Y., 1992, *J. Meas. and Cont.*, **31**, 473–478 (in Japanese).

134. Kelly, J., G. Leitmann, and A. Soldatos, 1987, *J. Opt. Theory and Appli.*, **53**, 159–180.

135. Fujita, T., Q. Feng, E. Takenaka, T. Takano, and Y. Suizu, 1988, *Proc. 9th World Conf. on Earthquake Engrg.*, Tokyo, Japan.

136. Pu, T.P. and J.M. Kelly, 1991, *J. Engrg. Mech. (ASCE)*, **117**, 2221–2236.

137. Inaudi, J., F. Lopez-Almansa, J.M. Kelly, and J. Rodellar, 1992, *Earthquake Engrg. and Struct. Dyn.*, **21**, 471–482.

138. Inaudi, J. and J.M. Kelly, 1990, *Proc. U.S. Nat. Workshop on Struct. Cont. Research*, edited by G.W. Housner and S.F. Masri, pp. 125–130.

139. Yang, J.N., A. Danielians, and S.C. Liu, 1991, *J. Engrg. Mech. (ASCE)*, **117**, 836–853.

140. Yang, J.N., A. Danielians, and S.C. Liu, 1992, *J. Engrg. Mech.* (*ASCE*), **118**, 1423–1456.
141. Yang, H.T.Y., D.G. Liaw, D.S. Hsu, and H.C. Fu, 1993, *J. Struct. Engrg.*, **119**, 902–919.
142. Yang, J.N., Z. Li, and S.C. Liu, 1992, *J. Engrg. Mech.* (*ASCE*), **118**, 1612–1630.
143. Yang, J.N., Z. Li, and S.C. Liu, 1992, *J. Engrg. Mech.* (*ASCE*), **118**, 2227–2245.
144. Clough, R.W. and J. Penzien, 1975, *Dyn. of Struct.* (McGraw Hill), New York.
145. Kerr, A.D. and M.A. El-Sibaie, 1987, *Earthquake Engrg. and Struct. Dyn.*, **15**, 549–563.
146. Crandall, S.H. and W.D. Mark, 1963, *Random Vibration in Mechanical Systems*, Academic Press.
147. Ayorinde, E.O. and G.B. Warburton, 1980, *Earthquake Engrg. and Struct. Dyn.*, **8**, 219–236.
148. Abdel-Rohman, M., 1984, *Building and Environment*, **19**, 191–195.
149. Kobori, T., N. Koshika, K. Yamada, and Y. Ikeda, 1991, *Earthquake Engrg. and Struct. Dyn.*, **20**, 135–149.
150. Kobori, T., N. Koshika, K. Yamada, and Y. Ikeda, 1991, *Earthquake Engrg. and Struct. Dyn.*, **20**, 151–166.
151. Ogata, K., 1990, *Modern Cont. Engrg.* (Prentice Hall), 2nd ed., Chap. 4, pp. 283–288.
152. Wylie, C.R. and L.C. Barrett, 1982, *Adv. Engrg. Math.* (McGraw-Hill), pp. 971–975.
153. Koizumi, T., Y. Furuishi, and N. Tsijiuchi, 1989, *Trans. of JSME*, **55**, 1602–1608 (in Japanese).
154. Tsijiuchi, N., T. Koizumi, and Y. Furuishi, 1991, *Trans. of JSME*, **57**, 63–68 (in Japanese).
155. Naiem, F., 1989, *The Seismic Design Handbook* (Van Nostrand Reinhold), edited by F. Naiem, New York.
156. Mondkar, D.P. and G.H. Powell, 1977, *Int. J. Num. Methods in Engrg.*, **11**, 499–520.
157. Gardiner, J.D., 1992, *J. Guidance, Cont. and Dyn.* (*AIAA*), **15**, 280–282.
158. Swei, S.M., C.C. Chang, and Y.T. Jiang, 1993, *Proc. Asia-Pacific Vib. Conf.*, Kitakyushu, Japan, pp. 84–89.
159. Tsai, H.C. and J.M. Kelly, 1989, *Earthquake Engrg. and Struct. Dyn.*, **18**, 551–564.
160. *Testline Mechanical Test System Components* (MTS System Co., 1987), Minneapolis, Minnesota.
161. Abdel-Rohman, M. and H.H. Leipholz,, 1983, *J. Struct. Engrg.* (*ASCE*), **109**, 628–645.
162. Davenport, A.G., 1967, *J. Struct. Division* (*ASCE*), **93**, 11–34.
163. Simiu, E., 1974, *J. Struct. Division* (*ASCE*), **100**, 1897–1910.
164. Shinozuka, M., C.B. Yun, and H. Seya, 1990, *J. Wind Engrg. and Indus. Aerodyn.*, **36**, 829–843.

4 BEHAVIOUR AND ANALYSIS TECHNIQUES OF TUBULAR BUILDINGS

Y. SINGH[1] and A.K. NAGPAL[2]

[1]*Department of Civil Engineering, Jamia Millia Islamia, New Delhi-110025, India*
[2]*Department of Civil Engineering, Indian Institute of Technology, Delhi, Hauz Khas, New Delhi-110016, India*

4.1. INTRODUCTION

Advent of tubular systems have been a significant development in the modern building structural systems. Most of the tallest buildings of the world both in steel as well as in concrete, have tubular systems as the basic load resisting system. The system not only provides high structural efficiency but also provides large column free space, suitable for flexible usage. The system derives its structural efficiency by having the most of the load carrying material on the periphery in the form of closely spaced columns interconnected by deep spandrel beams. The large number of rigidly connected joints provide a high lateral rigidity necessary to avoid premium for the heights achieved.

The basic form of the system, termed as framed tube, resembles in appearance and structural action, a hollow cantilever tube having perforations for windows (Figure 1). The system consists of four frame panels along the perimeter. The first variation of the basic form, to increase its efficiency is a tube-in-tube system (Figure 2) which consists of the outer framed-tube and interior shear wall core for services. The interior and the exterior tubes interact in a way similar to the frame-shear wall interaction.

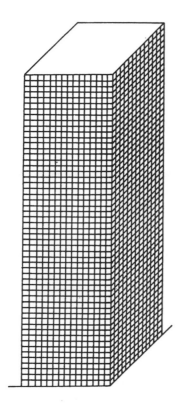

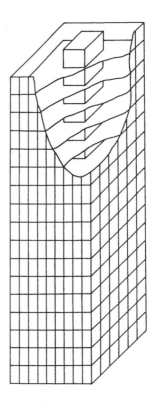

Figure 1. Framed tube.   **Figure 2.** Tube-in-tube.

The efficiency of a framed-tube system can be further increased by adding more number of frame panels in each orthogonal direction. The resulting system is termed as a multi-cell tube or a bundled tube (since a number of framed tubes are bundled together). In addition to the structural efficiency the system allows variation of the architectural form along the height (Figure 3). The system is notable in being used for the building which stands tallest in world till date (Sears Towers: 110 storeys, 442 m high).

Another way of increasing the efficiency of the basic system is by combining the frame and the truss actions in a braced-tube building. In steel buildings, this is achieved by providing a vertical truss in frame panels (Figure 4) and in concrete buildings by blocking the panels in a diagonal pattern (Figure 5).

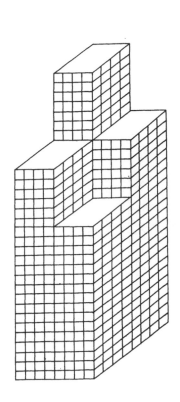

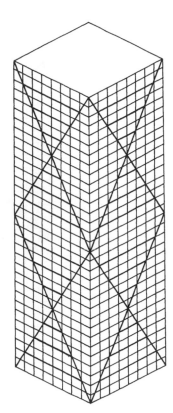

Figure 3. Multi-cell (bundled) tube.   **Figure 4.** Trussed tube (steel).

4.2. BEHAVIOUR

The primary loads acting on buildings, in general, can be broadly classified into two categories, namely, the lateral load and the vertical load. The vertical load originates from the gravity action while the lateral load originates from the action of wind or earthquake. The behaviour of the tubular buildings system is discussed, first under the lateral load and then under the vertical load.

4.2.1. Lateral Loads

Under the lateral loading, two modes of behaviour of a framed-tube system can be identified, viz. bending mode and shear mode. In bending mode,

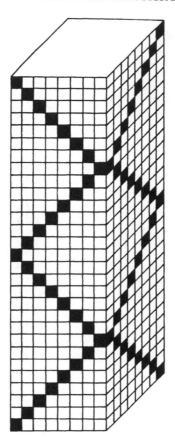

Figure 5. Trussed tube (concrete).

the overall overturning moment is resisted through the development of axial forces in the columns, while in the shear mode, storey shears are resisted through the development of shear forces primarily in web frames.

In the bending mode, interaction between the web and the flange frames takes place primarily through the transfer of vertical shear at the corners. A web frame in this mode may be visualized as a plane frame subjected to in-plane lateral load, and vertical shears acting at the two end column lines (Figure 6a) while the flange frame is to be visualized as a plane frame subjected to vertical shears along the column lines (Figure 6b).

As the beams interconnecting the columns have only finite shearing rigidity, the distribution of column axial forces is non-linear across the frame panels as shown in Figure 7. In the lower portion the corner column in the

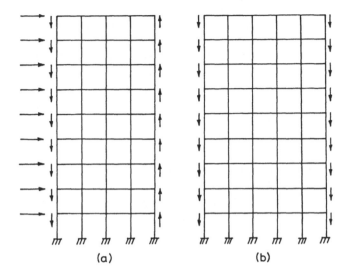

Figure 6. Frames under lateral loading: (a) Web frame; (b) Flange frame.

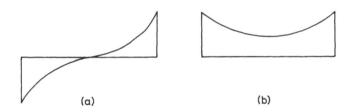

Figure 7. Distribution of column axial forces in the bottom portion of a framed-tube: (a) Web frame; (b) Flange frame.

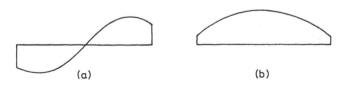

Figure 8. Distribution of column axial forces in the top portion of a framed-tube: (a) Web frame; (b) Flange frame.

flange tends to take more load than the interior columns. The phenomenon is termed as shear lag. This phenomenon has also been observed in box girders. However, in most of the top portion of the framed-tube the distribution of column axial forces is reverse (Figure 8) and then the phenomenon is termed as negative shear lag. Origin of the positive shear lag is obvious while that of the negative shear lag requires explanation. A physical explanation of the origin of the negative shear lag has been presented by the authors[1] elsewhere.

It has been observed that the shear lag behaviour in the flange frame of a framed tube is influenced by the type of lateral loading, in addition to the following structural parameters:

(i)   Stiffness Factor, S_f (Ratio of beam shearing stiffness to column axial stiffness),
(ii) Stiffness Ratio, S_r (Ratio of column flexural stiffness to beam flexural stiffness),
(iii) Height to weight ratio, λ.

The ratio, f of the axial force in the corner column to that in the central column can be chosen as a parameter which indicates the effect of shear lag. The typical variation of f along the height of a framed-tube under three types of lateral loading viz. uniformly distributed load, triangular load and point load, is shown in Figure 9. In the bottom portion, f is greater than unity indicating positive shear lag while in the top portion it is less than unity, indicating negative shear lag. The shear lag reversal from positive to negative takes place under all the three types of loadings. However, in case of the point load, the portion in which negative shear lag occurs is limited to only a few storeys at the top.

A detailed study of the effect of various structural parameters on the shear lag behaviour is available.[1] The shear lag, positive, as well as negative reduces with increase in S_f, S_r and λ.

In the shearing mode, the storey shears are resisted primarily through the inplane shearing action of the web frame and also through the out of plane bending of flange frames. The web frames and flange frames interact in a manner similar to the interaction in frame-shear wall systems; with the action of the web frame being similar to the frame action and that of the flange columns being similar to that of a weak shear wall. The interaction is shown to be secondary and has significant effect only in the bottom and the top portion of the building.[2]

In a tube-in-tube building an interaction similar to that of a frame-shear wall system takes place between the inner core and the outer framed tube. This results in a complex variation along the height, of the lateral load shared by the framed tube (Figure 10). As the shear lag behaviour of a framed-tube

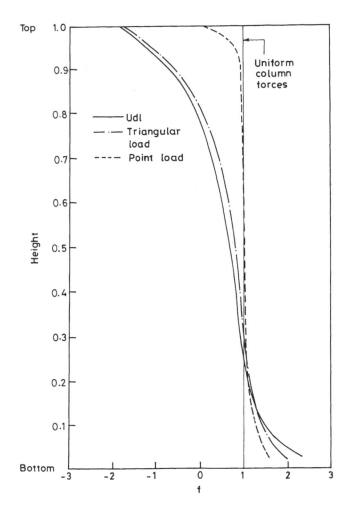

Figure 9. Variation of f with height for different types of loading.

is greatly influenced by the variation of lateral loading along the height, the resulting shear lag behaviour of the framed-tube in a tube-in-tube system is quite different from that of a framed-tube acting independently. Figure 11 shows typical variations of f along the height for different values of relative stiffness ρ (= ratio of flexural stiffness of the core to the sum of flexure stiffness of columns in web frames of the framed-tube). In the presence of the inner core the shear lag near the bottom and the top reduces considerably.

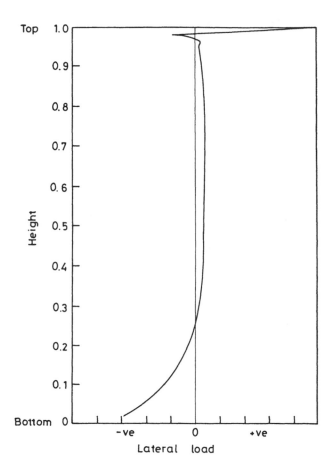

Figure 10. Variation of lateral load shared by the framed-tube along the height in a tube-in-tube building.

A multi-cell tube is a conglomerate of a number of framed tubes. The intermediate web frames tie up the flange frames and result in more uniform distribution of axial forces in flange frames than would occur in their absence. For a 2-cell-tube shown in Figure 12, the distribution of axial forces in flange columns is shown in Figure 13. Here two ratios are of interest: (1) ratio, f_1 of the axial force in the corner column to that in the mid-cell column and ratio f_2 of the axial force in the corner column to that in the end column of the interior web frame.

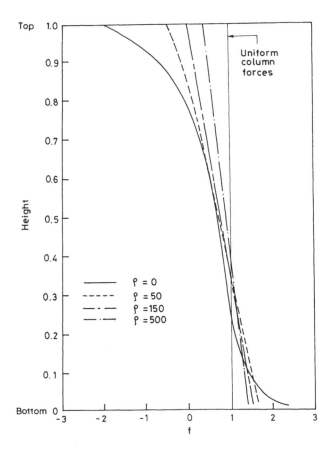

Figure 11. Variation of f with height for typical values of ρ in a tube-in-tube building.

The variation of the two ratios along the height of the building is shown in Figures 14 and 15 respectively, for different values of stiffness factor, S_f. The variation of f_1 is similar to that of f in a framed-tube while the variation of f_2 shows that the corner column is more stressed in the bottom portion while the end column of the interior web frame is more stressed in the top portion of the building.

In a braced-tube, in the bending mode, the effect of the bracing is to enhance the resistance of a frame panel against vertical shearing. This results in a more uniform distribution of column axial forces or in reduced shear lag effect.[3]

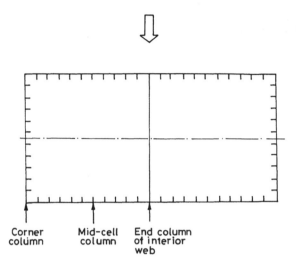

Figure 12. Schematic plan of 2-cell tube building.

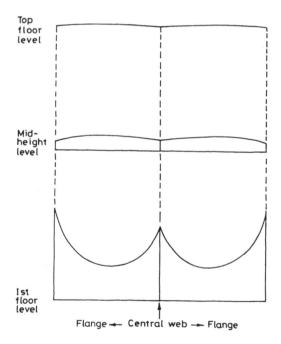

Figure 13. Distribution of axial forces in the flange columns of 2-cell tube building.

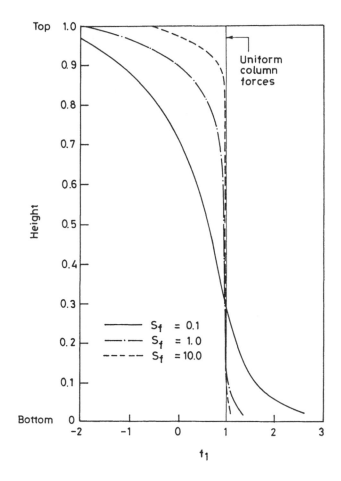

Figure 14. Variation of f_1 with height for typical values of S_f.

In the shearing mode the bracing resists the horizontal shear in conjunction with the web frame. This results in much reduced shears and bending moments in the web frame columns.[3,4]

4.2.2. Vertical Load

The nature of application of the two types of vertical loads (i.e. dead load and live load) on a building is quite different. The dead load builds up sequentially

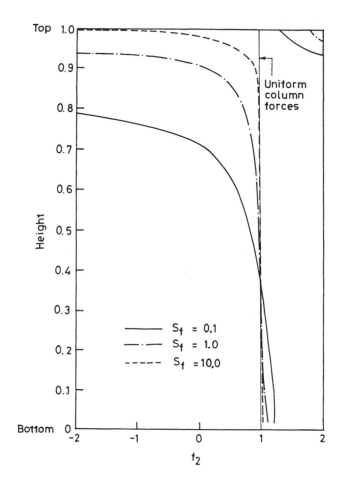

Figure 15. Variation of f_2 with height for typical values of S_f.

during construction, while the live load is applied after the construction of structure is completed. The dead load at any stage of construction is resisted only by the part of the structure completed at that stage, whereas the live load is resisted by the entire structure. In case of dead (sequential) load, the members in a storey are stressed due to loads acting on this storey and the storeys above. The loads acting on storeys below this storey do not affect the member forces. On the other hand, in case of live (simultaneous) load, the member forces in a storey are affected by the load on all the storeys.

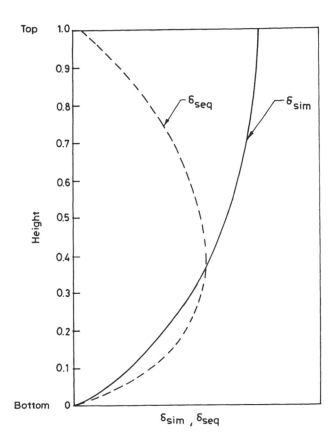

Figure 16. Variation of δ_{sim} and δ_{seq} along the height.

The differing behaviour under two types of vertical loads can be studied in terms of the variation along the height of differential shortening, δ, between the adjacent joints at a floor level. Typical variations of $\delta (= \delta_{sim}$ for simultaneous load, $= \delta_{seq}$ for sequential load) are shown in Figure 16. The parameter, $\delta_d = \delta_{sim} - \delta_{seq}$, to a large extent, is a measure of the difference in behaviour under simultaneous load and sequential load. The nature of variation of δ_d along the height is shown in Figure 17.

The effect of bracing in braced-tubes is similar to that of increase in shear stiffness of the beams. This results in more uniform distribution of column axial forces and reduction of differential shortening and hence reduction of beam end moments.

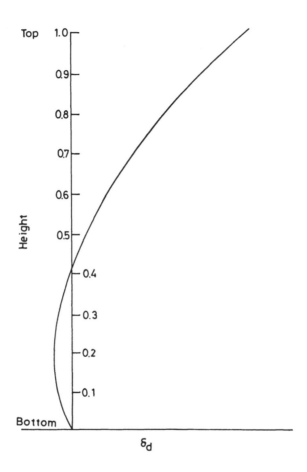

Figure 17. Variation of δ_d along the height.

4.3. ANALYSIS UNDER LATERAL LOAD

Two approaches for the analysis under lateral loads are available: (i) continuum approach, and (ii) discrete approach. In the continuum approach the frame panels are replaced by equivalent orthotropic plates. Closed-form solutions for uniform tubular systems for regular loadings are available. However, the modelling is approximate and solution for practical buildings having variation in member properties becomes cumbersome. The importance of the approach lies in obtaining rapid approximate solutions and in providing a general understanding of the behaviour of structures.

Stringers   Shear panels

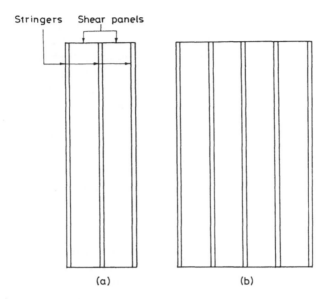

(a)   (b)

Figure 18. Stringer-shear panel model: (a) 3 Stringer model; (b) 5 Stringer model.

In the discrete approach the structure is modelled as an assemblage of discrete members. The approach is well suited for practical buildings.

A number of techniques based on the two approaches are available. A brief account of some important techniques is being given in the following sections.

4.3.1. Continuum Approach

This approach involves 2 basic steps: (1) replacement of a frame panel by an equivalent continuum and (2) solution of the continuum model. For the first step two modelling techniques are available. In technique 1[5-7] the frame panels are replaced by equivalent orthotropic plates, the properties of which are obtained by assuming points of contraflexure at suitable locations. In technique 2[8] the frame panels are replaced by a series of stringers and shear panels (Figure 18). A stringer can take only axial force and has no bending or shearing rigidity. The shear panel, on the other hand, can take only shear force and has no axial or bending rigidity.

Following technique 1, for the framed tube the elastic properties of the orthotropic plates are obtained by assuming points of contraflexure at mid-span of beams and columns.[5,6] Closed-form solutions for deflections and stresses are derived. The importance of shear deformation in members and of rigidity of joints has been pointed out and expressions for the elastic properties incorporating these aspects are available.[9] The technique has also been used for the torsional analysis of framed tubes[10] and closed-form solutions similar to those for bending have been obtained. In a variant of this procedure[7] the elastic properties are obtained by assuming different locations of points of contraflexure in different parts of the structure. Further the approximate deflections are obtained by Rayhigh-Ritz method. For a practical building with variations in member properties along the height for which closed-form solution cannot be obtained, a numerical technique based on fourth order Runge-Kutta method has been used.[11] The finite strip method has also been used for the solution of the model both for symmetric and asymmetric framed tube buildings.[12]

The technique has been extended to other forms of tubular buildings. In tube-in-tube buildings the outer tube is also converted into an equivalent continuum.[13] The compatibility between the equivalent outer continuum and the inner tube is enforced throughout the height. For the multi-cell tube also closed-form solutions have been obtained.[14] In trussed tubes the vertical truss is converted into an additional orthotropic plate having the same lateral rigidity as that of the truss.[15]

Limited work is available on use of technique 2. The frame-tube has been modelled[8] and the solution obtained using transfer matrix method. The technique has been employed by Kristek and Bauer[16] to estimate forces in columns on front faces of multistorey building. Solution is obtained using harmonic analysis.

Recently an analogy has been proposed[1] between a cantilever box girder and a framed-tube building employing which the height at which shear lag reversal takes places and forces in flange columns can be evaluated and is described in some detail here.

A procedure for estimating stress, σ_x and distance of shear lag reversal from the free end in cantilever box girder bridges has been developed[17] based on an earlier work[18] with some modifications in the nature of distribution of longitudinal displacements across the flange width. Expressions for longitudinal stress, σ_x and distance of reversal of shear lag, x_1 from the free end have been derived. Let Reissner's parameters k and n be expressed as

$$k = \frac{1}{b}\sqrt{\frac{14Gn}{5E}} \tag{1}$$

and

$$n = \frac{1}{1 - \frac{7I_s}{8I}} \tag{2}$$

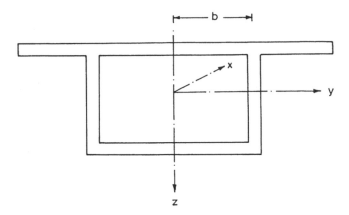

Figure 19. Cross-section of a box-girder.

in which 'b' is the half width of the flange of the box girder (Figure 19), E and G are Young's modulus and shear modulus respectively and I and I_s are moment of inertia of the entire cross-section and of the flange, respectively.

Expressions for σ_x and x_1 are then given by

$$\sigma_x = \pm q \frac{z}{I} \left\{ -x^2/2 + \left(1 - \frac{y^2}{b^2} - \frac{3I_s}{4I} \right) \frac{7n}{6k^2} \left[\frac{\cosh k(L-x) + kL \sinh kx}{\cosh kL} - 1 \right] \right\} \tag{3}$$

and

$$x_1 = \frac{1}{k} \sinh^{-1} \left[\frac{2kL \cosh kL - \sinh 2kL}{(kL)^2 - 2kL \sinh kL - 1} \right] \tag{4}$$

in which q is the uniformly distributed load, L, is the length of the box girder and x, y, z are the coordinates (Figure 19) of the point on the flange at which σ_x is evaluated.

Similar expressions for a triangular load and a point load are also available.[17]

The application of the above expressions for a uniform framed-tube consists in interpretation of the Reissner's parameters for framed-tube system. This is done by obtaining I and I_s by considering discrete column areas instead of the continuous distribution of material as in case of box girders.

The ratio G/E is interpreted as the ratio of the distributed shearing rigidity per unit height, S of a bay of the flange frame to the distributed axial rigidity K_a per unit width of the frame. Here it is important to note that the distribution of column axial forces in the flange is governed by the vertical shearing of a bay and not by the horizontal shearing of a storey, as assumed earlier.[6] The shearing rigidity of a bay of the flange frame can be obtained in a manner similar to that given by Basu and Nagpal[19] for shearing rigidity of a storey. The effect of restraint provided by columns against the rotation of beams can be modelled by equivalent springs[1] having stiffness K_s $(= 6I_c/H_s$, where I_c is the moment of inertia of column and H_s is the storey height). The expression for S can be obtained as

$$S = \frac{12EI_b}{H_s W_b^2}\left[\frac{\tau}{\tau + 6}\right] \tag{5}$$

where I_b and W_b are moment of inertia and span of beam and $\tau = K_s W_b/I_b$.
The expression for K_a can be written as

$$K_a = \frac{EA_c}{W_b} \tag{6}$$

where A_c is the area of the column.
Now the analogous Reissner's parameter k can be written as

$$k = \frac{1}{b}\sqrt{\frac{14SN}{5K_a}} \tag{7}$$

where b is the half width of the flange frame and n is obtained by considering discrete column areas.

Using analogous Reissner's parameters k and n, the longitudinal stress and the height of shear lag reversal can be obtained from Equations (3) and (4), respectively. Axial force in a column is obtained by multiplying the longitudinal stress, σ_x by the area of the column. It has been shown[1] that the analogy results in quite accurate estimate of column axial forces.

4.3.2. Discrete Approach

A tubular building is a three-dimensional assemblage of discrete skeletal members and can be modelled as a space frame. However, the computational effort required in the analysis of the space frame is high due to large number of degrees of freedom (DOFs). Procedures are available to reduce the computational effort.

Without much loss of accuracy, a building can be reduced to an equivalent building of reduced number of storeys by modifying the member properties.[20]

Symmetric tubular buildings can be converted into equivalent plane frames which yield quite accurate results. In an earlier work[21] the framed tube building was converted into an equivalent plane frame by neglecting the out of plan bending of flanges (Figure 20). The transfer of vertical shears between the web and the flange frames takes place through fictious horizontal members having infinite shearing rigidity but no axial and bending rigidities. Taking the advantage of symmetry only a quarter of the framed-tube need be considered.

As indicated earlier in Section 4.2.1, for the most of the height of the framed-tube the equivalent plane frame model yields accurate member forces. Only in the bottom and top portion of the building, the effect of out of plane bending of flange frame is significant. This results in erroneous column moments and shear forces in a few bottom and top storeys.[3,22] It has been suggested[22] that this out-of-plane bending of the flange frame can also be incorporated in the equivalent plane from model, in the form of an additional cantilever column (Figure 21) interconnected with the web frame by rigid links at every floor level. The modified model has been shown[3] to yield accurate member forces, throughout the height of the building.

The equivalent plane frame modelling has been extended for multi-cell tube systems as well.[23] The orthogonal frame panels are put in a single plane in series and the compatibility of vertical and horizontal displacements is enforced either through fictitious members or by inter-nodal constraining. Figure 22 shows the schematic representation of the modified equivalent plane-frame models for tubes with different number of cells, with appropriate boundary conditions for enforcing symmetry and fictitious members for enforcing compatibility. The accuracy of the model has been studied[24] by comparing the results with those obtained from the 3-D analysis. The model yields quite accurate results throughout the height of the building.

The equivalent plane frame model has also been used for estimating the free vibration characteristics.[25] The effect of joint rigidity on free vibration characteristics of a framed-tube has been studied[26] by comparing the analytical results with those obtained experimentally. The joint rigidity does not have a significant effect on mode shapes but it reduces the time periods and should be considered in the analytical model.

An alternative solution procedure based on the equivalent plane frame idealization has been proposed by Ast and Schwaighofer.[27] In this procedure, the web and the flange panels are visualized to be separated. The interaction between them is considered by satisfying the compatibility of vertical deformations at floor levels. It has been shown that the computational effort can be reduced by satisfying the compatibility at a reduced number of levels.

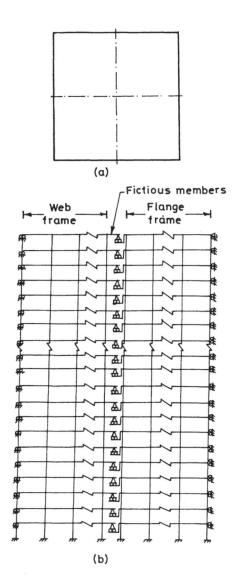

(a)

(b)

Figure 20. Equivalent plane frame model of a symmetric framed-tube building: (a) Symmetric building; (b) Plane frame model.

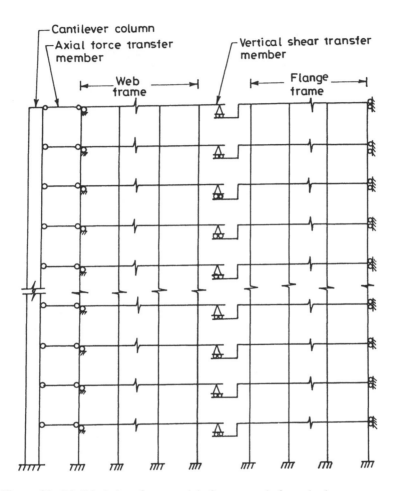

Figure 21. Modified plane frame model of a symmetric framed-tube.

The computational effort required and the accuracy of results is proportional to the number of levels at which the compatibility is enforced.

To reduce the computational effort, Declereq and Powell[28] proposed a macro-element model. The model reduces the number of DOFs and hence the computational effort. A macro-element contains a number of beam and column members and embraces a homogeneous portion of a frame panel. Contribution of an individual member to the stiffness of the corresponding macro-element is obtained through use of shape functions. The model can also be used for framed-tubes of arbitrary plan layouts by considering the

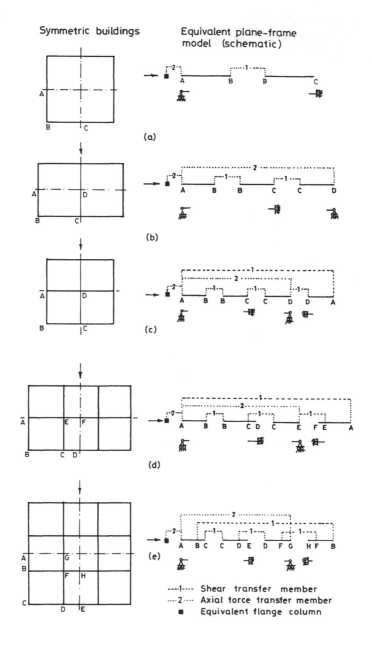

Figure 22. Schematic representation of the plane-frame models for multi-cell tubes: (a) 1-Cell; (b) 2-Cell; (c) 4-Cell; (d) 6-Cell; (e) 9-Cell.

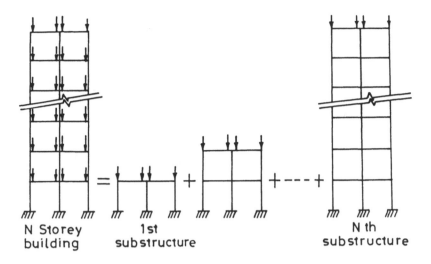

Figure 23. Sequential analysis of N-storey building.

corner columns as separate macro-elements. The model yields approximate results suitable for preliminary design with much reduced computational effort.

4.4. ANALYSIS UNDER VERTICAL LOAD

The development of simplified techniques for vertical load analysis of tubular buildings has not received much attention. Owing to a large number of storeys, considerable differential column shortening gets accumulated along the height. Therefore, consideration of nature of loading — sequential or simultaneous becomes important.

For sequential load the analysis of a 'N' storey building can be visualized as the analysis of N substructures having number of storeys varying from 1 to N (Figure 23). The final member forces are the sum of the member forces obtained from the analysis of individual substructures.

For simultaneous load, on the other hand, only one analysis of the complete building loaded at all the floors is required.

Limited work in recent past has been done on the sequential load analysis of framed buildings which should be applicable for tubular buildings also. Procedures in which one floor or a number of floors at a time are considered to be loaded are available.[29,30]

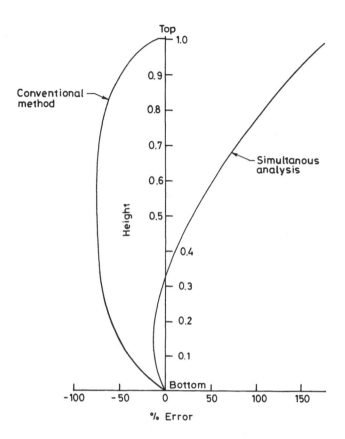

Figure 24. Typical variations of % errors in end beam moments.

In the design office, it is usual to carry out analysis for dead load also assuming it be simultaneously applied at floor levels of the completed structure as in the case of live load. This simultaneous analysis for dead load causes error in member forces which is primarily due to difference, $\delta_d (= \delta_{\text{sim}} - \delta_{\text{seq}})$. Figure 24 shows a typical variation of error in end moment in beams. The error is the maximum in the top portion of the building.

A correction factor method for framed building has been given[30] in which the correction δ_d to be applied to δ_{sim} is estimated using design curves for normalized δ_d. These design charts have been prepared considering a few practical buildings.

The corrections for end moment, M and shear, S to be applied to these quantities obtained from simultaneous analysis for a beam of length, L and flexural rigidity, EI are given simply as

$$M = 6EI\delta_d/L^2 \tag{8}$$

$$S = 12EI\delta_d/L^3 \tag{9}$$

The method has been found to be effective particularly for flexible buildings. Similar design curves need to be developed for tubular buildings.

In the absence of availability of specialized softwares for sequential analysis it has also been recommended[31] that the conventional method in which one floor at a time is considered, may be preferred over the simultaneous analysis. In a recent study on frames[32] the errors in bending moments obtained from the conventional method and the simultaneous analysis have been evaluated for a number of frames. A typical variation of errors in beam end moments obtained from conventional analysis is shown in Figure 24. The error is the maximum in the middle portion of the frame. It is shown in the study that neither the simultaneous analysis nor the conventional procedure yields sufficiently accurate results throughout the height of the building and that except for the top portion of the building, the conventional method results in greater error than the simultaneous analysis in many practical buildings.

4.5. TWO-STAGE SOLUTION PROCEDURE

Recently a two-stage solution procedure[33-35] for the analysis of tubular buildings has been proposed. The procedure reduces the computational effort and computer storage locations drastically. The solution is obtained in two stages. In stage 1, an approximate solution is obtained using one or a combination of simplified models/approximate analysis procedures. In stage 2, the approximate solution is iteratively refined to yield accurate solution. The CPU time and computer storage locations required are a fraction of those required in the standard single stage procedure. Further, the procedure is synchronous with the total design process (Figure 25). At the preliminary design stage, depending on the accuracy desired, the solution of stage 1 itself (or with some refinement in stage 2) can be used, while at the final design stage the converged solution of stage 2 is used.

The concept is quite general and is applicable to other large structural systems as well.

For lateral load analysis and estimation of free vibration characteristics of framed tubes, the stage 1 solution is obtained by constraining DOFs[33,34] in the plane frame model. The building is divided into a number of segments, each segment spanning across a number of storeys; Figure 26 shows the

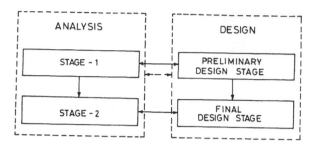

Figure 25. Two-stage analysis – design procedure.

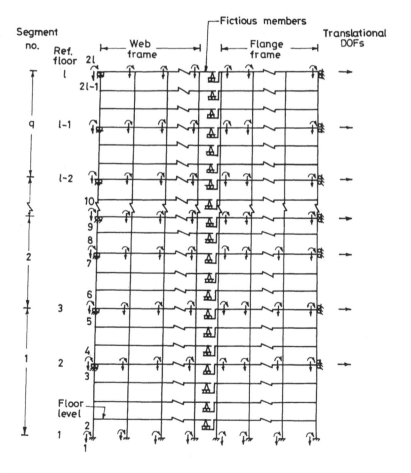

Figure 26. Constrained plane-frame model of a symmetric framed-tube under lateral loading.

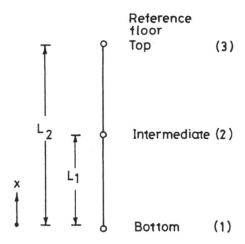

Figure 27. Reference floors and levels of DOFs in a segment.

division into 'q' number of segments. Within each segment three reference floors are defined, one each at the bottom floor level and the top floor level and the remaining at an intermediate floor (Figure 27). The total number of reference floors, 'l' in the building is much smaller than the number of floors (Figure 26). Two levels of DOFs are identified. Level 1 DOFs, as usual, comprise of a translational DOF at each floor level and a vertical and a rotational DOF at every joint of floor levels. Similar Level 2 DOFs are identified at selected reference floors only and are shown in Figure 26. Level 1 and Level 2 DOFs are assumed to be related as

$$u^1 = \sum_{j=1}^{3} N_j u_j^2 \tag{10a}$$

$$v^1 = \sum_{j=1}^{3} N_j v_j^2 \tag{10b}$$

$$\theta^1 = \sum_{j=1}^{3} N_j \theta_j^2 \tag{10c}$$

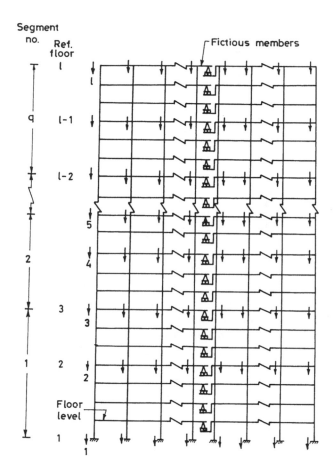

Figure 28. Constrained plane-frame model of a symmetric framed-tube under vertical loading.

where u^1, v^1, θ^1 are level 1 DOFs (lateral displacement, vertical displacement and rotation, respectively, at a joint of a floor), u_j^2, v_j^2, θ_j^2 are the corresponding level 2 DOFs at the jth reference floor of the segment (Figure 27) and N_j are the shape functions given as

$$N_1 = \frac{(L_1 - x)(L_2 - x)}{L_1 L_2} \tag{11a}$$

$$N_2 = \frac{x(L_2 - x)}{L_1(L_2 - L_1)} \tag{11b}$$

$$N_3 = \frac{x(x - L_1)}{L_2(L_2 - L_1)} \tag{11c}$$

where further x is the distance of a floor level from the bottom reference floor.

Stage 1 results in an auxiliary problem of much reduced size and provides an estimate of the deflections and member forces.

In stage 2, the constraints applied in stage 1 are relaxed using an iterative procedure such as Gauss Seidel procedure with Successive Over-Relaxation or Preconditioned Conjugate Gradient scheme.

In case of free vibration characteristics[34] the flexibility matrix of constrained structure is obtained in stage 1 any of the iterative procedures mentioned above is then used to remove the constraints. The resulting flexibility matrix is used to solve the eigen-problem to get frequencies and mode-shapes of the framed-tube.

In case of vertical loading, the nature of constraints to be imposed on rotational DOFs is not apparent since relative values of rotations at joints are dependent on the loading pattern. Therefore, rotational DOFs are delinked from vertical DOFs and stage 1 itself obtained in two steps.[35] In step 1, for the estimation of vertical displacements, a constrained plane frame model (Figure 28) similar to that for lateral load but having only vertical DOFs is used. The beams are accordingly replaced by equivalent shear beams having one translational DOF at each end.

In step 2, joint rotations are obtained using isolated storey substructures. The substructures are analyzed recursively from the bottom to the top considering known vertical displacements and rotations at the lower floor level from step 1, along with the applied loading (Figure 29).

Stage 1 results in a good estimate of joint displacements and rotations which are refined in stage 2 to yield accurate deflections and member forces.

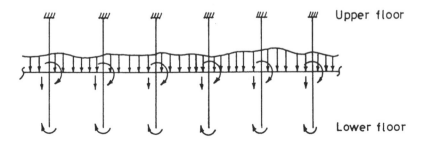

Figure 29. Segments of a substructure for evaluation of joint rotations at a floor.

References

1. Singh, Y. and A.K. Nagpal, 1994, *J. Struct. Engg.*, ASCE, **120**(11), 3105–3121.
2. Singh, Y. and A.K. Nagpal, 1993, *Struct. Design, Tall Buildings*, **2**, 325–331.
3. Stafford Smith, B. and A. Coull, 1991, *Tall Building Structures: Analysis and Design* (John Wiley & Sons, Inc., New York), Chap. 12, pp. 283–307.
4. Grossman, J.S., M. Cruvellier and B. Stafford Smith, 1986, *Concrete Int.*, **8**(9), 32–42.
5. Coull, A. and A.K. Ahmed, 1970, *J. Struct. Div., ASCE*, **104**(5), 857–862.
6. Coull, A. and B. Bose, 1975, *J. Struct. Div., ASCE*, **101**(11), 2223–2240.
7. Chang, P.C. and D.A. Foutch, 1984, *J. Struct. Div., ASCE*, **110**(12), 2955–2975.
8. Connor, J.J. and C.C. Pouangare, 1991, *J. Struct. Div., ASCE*, **117**(12), 3623–3644.
9. Ahmed, A.K. 1979, Ph.D. Thesis, University of Strathelyde, Glasgow, Scotland.
10. Coull, A. and B. Bose, 1976, *J. Struct. Div., ASCE*, **102**(12), 2366–2370.
11. Chang, P. and D. Foutch, 1985, *Computs. and Structs.*, **21**(4), 771–776.
12. Han, P.S., 1989, *Appl. Math. Modelling*, **13**, 348–356.
13. Chang, P.C., 1985, *J. Struct. Div., ASCE*, **111**(6), 1326–1337.
14. Coull, A., B. Bose and A.K. Ahmed, 1982, *J. Struct. Div., ASCE*, **108**(5), 1140–1153.
15. Chang, P., 1982, Ph.D. Thesis, Graduate College of the University of Illinois at Urbana Champaign.
16. Kristek, V. and K. Bauer, 1993, *J. Struct. Div., ASCE*, **119**(5), 1464–1483.
17. Chang, S.T. and F.Z. Zheng, 1987, *J. Struct. Div., ASCE*, **113**(1), 20–35.
18. E. Reissner, 1946, *Quarterly of Appl. Math.*, **4**(3), 268–278.
19. Basu, A.K. and A.K. Nagpal, 1980, *J. Struct. Div., ASCE*, **106**(5), 1175–1190.
20. Khan, F.R. and N.R. Amin, 1972, *The Structural Engineer*, **51**, 613–632.
21. Coull, A. and N.K. Subedi, 1971, *J. Struct. Div., ASCE*, **97**(8), 2097–2105.
22. Rutenberg, A., 1972, *J. Struct. Div., ASCE*, **98**, 942–943.
23. Stafford Smith, B., A. Coull and M. Cruvellier, 1988, *Computs. and Structs.*, **29**(2), 257–263.
24. Singh, Y., 1994, Ph.D. Thesis, Indian Institute of Technology, New Delhi, India.
25. Anderson, J.C. and G. Gurfinkel, 1975, *Earthquake Engineering, Struct. Dyn.*, **4**, 145–162.
26. Maison, B.F. and C.F. Neuss, 1985, *J. Struct. Div., ASCE*, **111**(7), 1539–1572.
27. Ast, P.F. and J. Schwaighofer, 1974, *Build, Sci.*, **9**, 73–77.
28. Declereq, H. and G.H. Powell, 1976, Report No., EERC 76–5, Earthquake Engg. Res. Centre, University of California, Berkeley, California.
29. Saffarini, H.S. and E.L. Wilson, 1983, UCB/SEMM 83/08, Deptt. of Civil Engg., Univ. of California, Berkeley, Calif.
30. Chang-Koon Choi and E-Doo Kim, 1985, *J. Struct. Div., ASCE*, **111**(11), 2373–2384.
31. Fintel, M., 1986, "Multistorey Structures", *Handbook of Concrete Engineering*, M. Fintel, ed., CBS Publishers and Distributors, New Delhi, India, Chap. 10, pp. 339–383.
32. Singh, Y., R. Goel and A.K. Nagpal, 1995, *Proc. National Seminar on High Rise Structures: Design and Construction Practices for Middle Level Cities*, The Institution of Engineers, Allahabad, India: II-13–22.
33. Singh, Y. and A.K. Nagpal, 1994, *Computs. and Structs.*, **50**(5), 655–663.
34. Singh, Y. and A.K. Nagpal, 1994, *Struct. Design Tall Buildings*, **3**, 37–49.
35. Singh, Y. and A.K. Nagpal, 1994, *Struct. Design Tall Buildings*, 65–83.

5 TECHNIQUES IN COMBINED BOUNDARY ELEMENT / FINITE STRIP / FINITE ELEMENT ANALYSIS OF BRIDGES

M.S. CHEUNG and W. LI

Department of Civil Engineering, University of Ottawa, Ottawa, Ontario, Canada K1N 9B4

5.1. INTRODUCTION

If a plate structure has constant cross-section and its end support condition does not change transversely, the finite strip method has proven to be a very efficient numerical structural analysis method.[1,2] However, if the structure has any irregularities, e.g. a rectangular plate with openings, the finite strip method is no longer applicable on its own and the finite element method or the boundary element method has to be used. In this case, however, if these methods can be combined together, with the finite strips being used for the regular part of the plate and the finite elements or boundary elements modelling the irregular part, then the efficiency of the finite strip method and the universality of the latter methods are both utilized to their full advantage.[3,4] Another case favorable for the combined analysis is a slab-on-girder bridge or box-girder bridge under the moving wheel loading. If the local effect of the load is concerned, the semi-analytical finite strip method becomes uneconomical because too many series terms must be used. In this situation, the spline finite strip method may give better efficiency. However, in order to obtain accurate results for the maximum bending moments, the load application point must be taken as a node and a dense mesh is required around this point. When the load is moving along the bridge, the mesh must be modified for each new position of the load. This problem can be solved by using boundary element method to analyze the slab, because only the boundary of the slab panel is divided into elements and there is no any mesh inside the panel. Consequently, no mesh modification is necessary while the load is moving. In this case, the girders are still analyzed by the finite strip method since only their deformation, not stresses, have significant influence on the local bending moments in the slab.[5]

In this application, special transition elements are used to connect the two different regions. One side of such a transition element coincides with the nodal line of adjacent finite strip and has the same degrees of freedom; on the opposite side there are a number of nodes which are connected with the finite element or boundary element. Inside the transition element the deflection is expressed in terms of the degrees of freedom of the finite strip nodal line and the nodes of the finite element or boundary element by their corresponding shape functions. Using the principle of minimum total potential energy, the stiffness matrix and load vector of the transition element can be obtained.

The girder can also be modelled by finite elements and the combined finite element/boundary element analysis is performed. This approach is applicable to more complicated girders such as the girders with openings or variable depth.

This chapter presents the basic procedures in

(1) Finite strip method for analysis of plates and folded plates.
(2) Combined finite strip/finite element analysis of irregular plates.
(3) Combined finite strip/boundary element analysis of irregular plates.
(4) Combined finite strip/boundary element analysis of slab girder bridges.
(5) Combined finite element/boundary element analysis of box girder bridges.

Numerical examples are given to show the applicability of above approaches.

5.2. FINITE STRIP METHOD FOR ANALYSIS OF PLATES

5.2.1. Displacement Function

Figure 1 illustrates a simply supported rectangular plate under transverse loading. In order to evaluate deflection and stresses, the plate can be divided into a number of longitudinal strips (Figure 2), which are smoothly connected to each other along strip boundaries referred to as nodal lines, and all the displacement fields of the individual strips constitute the deformation pattern of the entire plate. Since the ends of the strip are simply supported, the

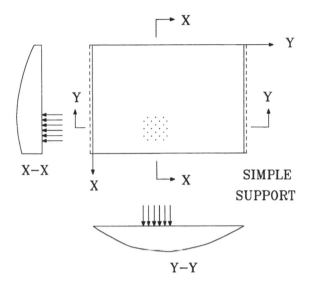

Figure 1. Plate in bending.

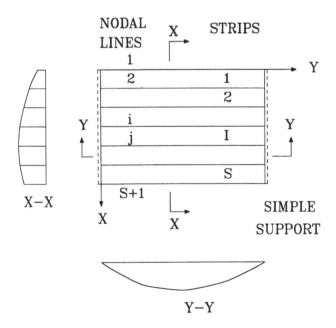

Figure 2. Plate divided into strips.

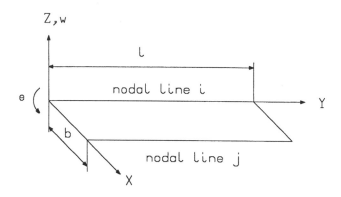

Figure 3. Plate strip.

appropriate displacement function must give zero deflection at both ends. Furthermore, the longitudinal bending moments must also be zero at the support sections. A series of sinusoidal functions can be taken to represent the deformed surface in the longitudinal direction, and a cubic polynomial in each strip may produce adequate simulation for the deflected shape in the transverse direction. Thus, the deflection w within a strip (Figure 3) can be expressed in terms of displacement parameters of nodal lines as follows:

$$w(x, y) = \sum_{m=1}^{r} (N_1(x)w_{im} + N_2(x)\theta_{im} + N_3(x)w_{jm} + N_4(x)\theta_{jm}) \sin \frac{m\pi y}{l}$$

$$(1)$$

where w_{im} and w_{jm} are the deflection amplitudes at the nodal lines i and j, respectively; θ_{im} and θ_{jm} represent the slope $(\frac{\partial w}{\partial x})$ amplitudes at these two nodal lines;

$$N_1(x) = (1 - 3X^2 + 2X^3)$$
$$N_2(x) = x(1 - 2X + X^2)$$
$$N_3(x) = (3X^2 - 2X^3)$$
$$N_4(x) = x(X^2 - X)$$
$$(X = x/b)$$

$$(2)$$

are the Hermitian polynomials; l and b are respectively the length and width of the strip; and r is number of series terms required in the analysis.

It can be seen that Equation 1 yields zero deflection and zero longitudinal curvature at two ends of the strip. It can also be shown that the displacement function in Equation 1 satisfies the continuity conditions across the strip boundary, since two adjacent strips connected along nodal line i have the same deflection and the same slops at any point on this nodal line as below:

$$w_i = \sum_{m=1}^{r} w_{im} \sin \frac{m\pi y}{l}$$

$$\left(\frac{\partial w}{\partial x}\right)_i = \sum_{m=1}^{r} \theta_{im} \sin \frac{m\pi y}{l}$$

$$\left(\frac{\partial w}{\partial y}\right)_i = \sum_{m=1}^{r} w_{im} \frac{m\pi}{l} \cos \frac{m\pi y}{l}$$

Equation 1 can be rewritten in a matrix form:

$$w(x, y) = \sum_{m=1}^{r} [N_1, N_2, N_3, N_4] \begin{Bmatrix} w_{im} \\ \theta_{im} \\ w_{jm} \\ \theta_{jm} \end{Bmatrix} \sin \frac{m\pi y}{l} \qquad (3)$$

or concisely

$$w(x, y) = \sum_{m=1}^{r} [N]\{\delta\}_m \sin \frac{m\pi y}{l} \qquad (4)$$

where $\{\delta\}_m$ includes the displacement parameters and $[N]$ is a matrix of transverse shape functions.

5.2.2. Energy Formulation

The strain energy of a plate strip is given as below[6]:

$$U = \frac{1}{2} \int_0^l \int_0^b \left(M_x \frac{\partial^2 w}{\partial x^2} + M_y \frac{\partial^2 w}{\partial y^2} - 2M_{xy} \frac{\partial^2 w}{\partial x \partial y} \right) dxdy \qquad (5)$$

where M_x, M_y and M_{xy} are the transverse bending moment, the longitudinal bending moment and the twisting moment, respectively. Equation 5 can be written in the following matrix form:

$$U = \frac{1}{2} \int_0^l \int_0^b [M_x, M_y, M_{xy}] \begin{Bmatrix} \frac{\partial^2 w}{\partial x^2} \\ \frac{\partial^2 w}{\partial x^2} \\ -2\frac{\partial^2 w}{\partial x \partial y} \end{Bmatrix} dxdy = \frac{1}{2} \int_0^l \int_0^b \{M\}^T \{\kappa\} dxdy$$

$$(6)$$

in which $\{M\}$ is the moment vector and $\{\kappa\}$ is the curvature vector.

From Equation 4, the curvature vector can be expressed in terms of the displacement parameters as:

$$\{\kappa\} = \left[\frac{\partial^2 w}{\partial x^2}, \frac{\partial^2 w}{\partial y^2}, -2\frac{\partial^2 w}{\partial x \partial y}\right]^T = \sum_{m=1}^{r}[B]_m\{\delta\}_m \tag{7}$$

where $[B]_m$ is the strain matrix in the following form:

$$[B]_m =$$

$$\begin{bmatrix} N_1'' \sin k_m y & N_2'' \sin k_m y & N_3'' \sin k_m y & N_4'' \sin k_m y \\ -k_m^2 N_1 \sin k_m y & -k_m^2 N_2 \sin k_m y & -k_m^2 N_3 \sin k_m y & -k_m^2 N_4 \sin k_m y \\ -2k_m N_1' \cos k_m y & -2k_m N_2' \cos k_m y & -2k_m N_3' \cos k_m y & -2k_m N_4' \cos k_m y \end{bmatrix} \tag{8}$$

in which $N' = \frac{dN}{dx}$, $N'' = \frac{d^2N}{dx^2}$ and $k_m = m\pi/l$.

The relationships between the moments and the curvatures are as follows:

$$M_x = D_x \frac{\partial^2 w}{\partial x^2} + D_1 \frac{\partial^2 w}{\partial y^2}$$

$$M_y = D_y \frac{\partial^2 w}{\partial y^2} + D_1 \frac{\partial^2 w}{\partial x^2} \tag{9}$$

$$M_{xy} = -2D_{xy} \frac{\partial^2 w}{\partial x \partial y}$$

where D_x and D_y are the flexural rigidities, D_{xy} is the torsional rigidity, and D_1 is the coupling rigidity. They are defined as

$$D_x = \frac{E_x t^3}{12(1 - \nu_x \nu_y)}$$

$$D_y = \frac{E_y t^3}{12(1 - \nu_x \nu_y)} \tag{10}$$

$$D_{xy} = \frac{E_{xy} t^3}{12}$$

$$D_1 = \nu_x D_x = \nu_y D_y$$

in which E_x, E_y and E_{xy} are the elastic moduli;

ν_x is the Poisson's ratio in the x direction, i.e. $\epsilon_x = \frac{\nu_x}{E_y}\sigma_y$ if $\sigma_x = \sigma_z = 0$;

ν_y is the Poisson' ratio in the y direction; and $\nu_y = \nu_x E_x/E_y$;

t is the thickness of plate strip.

For an isotropic plate, it is only necessary to set $E_x = E_y = E$, $\nu_x = \nu_y = \nu$, and $E_{xy} = G = E/2(1 + \nu)$.

In a matrix form, Equation 9 becomes

$$\{M\} = \left\{ \begin{array}{c} M_x \\ M_y \\ M_{xy} \end{array} \right\} = \left[\begin{array}{ccc} D_x & D_1 & 0 \\ D_1 & D_y & 0 \\ 0 & 0 & D_{xy} \end{array} \right] \{\kappa\} = [D]\{\kappa\} \qquad (11)$$

where $[D]$ is referred to as the elasticity matrix.

Substitution of Equation 7 into Equation 11 yields

$$\{M\} = \sum_{m=1}^{r} [D][B]_m \{\delta\}_m \qquad (12)$$

Then, by substituting Equation 7 and 12 into Equation 6, the strain energy of the strip can be expressed in terms of the displacement parameters as follows:

$$U = \frac{1}{2} \sum_{m=1}^{r} \sum_{n=1}^{r} \{\delta\}_m^T \int_0^l \int_0^b [B]_m^T [D][B]_n dxdy \{\delta\}_n \qquad (13)$$

It can be seen from Equation 8 that Equation 13 contains the following integrations:

$$\int_0^l \sin \frac{m\pi y}{l} \sin \frac{n\pi y}{l} dy = \int_0^l \cos \frac{m\pi y}{l} \cos \frac{n\pi y}{l} dy = \left\{ \begin{array}{ll} \frac{l}{2} & \text{if } m = n \neq 0 \\ 0 & \text{if } m \neq n \end{array} \right.$$
$$(14)$$

Because of the orthogonal properties of the trigonometrical functions, Equation 13 can be simplified as

$$U = \frac{1}{2} \sum_{m=1}^{r} \{\delta\}_m^T \int_0^l \int_0^b [B]_m^T [D][B]_m dxdy \{\delta\}_m \qquad (15)$$

For the plate strip under distributed transverse load of intensity $q(x, y)$, the potential energy of external loading is defined as

$$W = -\int_0^l \int_0^b q(x, y) w dxdy \qquad (16)$$

By substituting displacement function in Equation 4 into above expression, the potential energy of external loading can also be expressed in terms of the displacement parameters as below:

$$W = -\sum_{m=1}^{r} \{\delta\}_m^T \int_0^l \int_0^b [N]^T q(x, y) \sin \frac{m\pi y}{l} dxdy \qquad (17)$$

5.2.3. Stiffness Matrix and Load Vector

After integration, Equation 15 and Equation 17 are reduced to:

$$U = \frac{1}{2} \sum_{m=1}^{r} \{\delta\}_m^T [k]_m \{\delta\}_m \tag{18}$$

$$W = - \sum_{m=1}^{r} \{\delta\}_m^T \{p\}_m \tag{19}$$

where $[k]_m$, a 4×4 matrix, and $\{p\}_m$, a 4×1 vector, are the stiffness matrix and the load vector of the strip, respectively. Their entries are listed in Tables 1 and 2.

The strain energy of the entire plate and the potential energy of the external loads on the whole structure are obtained by summing up the contributions from all the strips:

$$U_t = \frac{1}{2} \sum_{I=1}^{S} \sum_{m=1}^{r} \{\delta\}_{Im}^T [k]_{Im} \{\delta\}_{Im} \tag{20}$$

$$W_t = - \sum_{I=1}^{S} \sum_{m=1}^{r} \{\delta\}_{Im}^T \{p\}_{Im} \tag{21}$$

where S is the number of strips into which the plate is divided, I is the subscript identifying an individual strip, and t is the subscript representing the whole structure.

Exchanging the sequence of summations in Equation 20 and Equation 21 yields

$$U_t = \frac{1}{2} \sum_{m=1}^{r} \sum_{I=1}^{S} \{\delta\}_{Im}^T [k]_{Im} \{\delta\}_{Im} \tag{22}$$

$$W_t = - \sum_{m=1}^{r} \sum_{I=1}^{S} \{\delta\}_{Im}^T \{p\}_{Im} \tag{23}$$

In order to carry out the summation over all the strips, a so-called assembling procedure is required.

First, for the m-th series term, the displacement parameters of the entire plate may be written as a $N \times 1$ vector $\{\delta\}_{tm}$, with N being the total number of these parameters. Next, the stiffness matrix $[k]_{Im}$ and the load vector $\{p\}_{Im}$ of strip I can be expanded into a $N \times N$ square matrix $[K]_{Im}$ and

Table 1. Stiffness matrix of plate strip

$$[k]_m = \begin{bmatrix} [K_{ii}]_b & [K_{ij}]_b \\ [K_{ij}]_b^T & [K_{jj}]_b \end{bmatrix} = \begin{bmatrix} k_1 & & \text{symmetrical} & \\ k_3 & k_2 & & \\ k_4 & k_5 & k_1 & \\ -k_5 & k_6 & -k_3 & k_2 \end{bmatrix}$$

where

$$k_1 = \frac{13lb}{70} k_m^4 D_y + \frac{12l}{5b} k_m^2 D_{xy} + \frac{6l}{5b} k_m^2 D_1 + \frac{6l}{b^3} D_x$$

$$k_2 = \frac{lb^3}{210} k_m^4 D_y + \frac{4lb}{15} k_m^2 D_{xy} + \frac{2lb}{15} k_m^2 D_1 + \frac{2l}{b} D_x$$

$$k_3 = \frac{11lb^2}{420} k_m^4 D_y + \frac{l}{5} k_m^2 D_{xy} + \frac{3l}{5} k_m^2 D_1 + \frac{3l}{b^2} D_x$$

$$k_4 = \frac{9lb}{140} k_m^4 D_y - \frac{12l}{5b} k_m^2 D_{xy} - \frac{6l}{5b} k_m^2 D_1 - \frac{6l}{b^3} D_x$$

$$k_5 = \frac{13lb^2}{840} k_m^4 D_y - \frac{l}{5} k_m^2 D_{xy} - \frac{l}{10} k_m^2 D_1 - \frac{3l}{b^2} D_x$$

$$k_6 = -\frac{lb^3}{280} k_m^4 D_y - \frac{lb}{15} k_m^2 D_{xy} - \frac{lb}{30} k_m^2 D_1 + \frac{l}{b} D_x$$

Table 2. Load vector of plate strip

1. Point force P_o at (x_o, y_o)

$$\{p\}_m = \begin{Bmatrix} Z_{im} \\ M_{im} \\ Z_{jm} \\ M_{jm} \end{Bmatrix} = \begin{Bmatrix} N_1(x_o) \\ N_2(x_o) \\ N_3(x_o) \\ N_4(x_o) \end{Bmatrix} P_o \sin k_m y_o$$

2. Uniform load of intensity Q_o on the entire strip

$$\{p\}_m = \begin{Bmatrix} Z_{im} \\ M_{im} \\ Z_{jm} \\ M_{jm} \end{Bmatrix} = \begin{Bmatrix} \frac{b}{2} \\ \frac{b^2}{12} \\ \frac{b}{2} \\ -\frac{b^2}{12} \end{Bmatrix} [1 - (-1)^m] \frac{Q_o l}{m\pi}$$

3. Patch load of intensity Q_o on area $[x_1, x_2]$ and $[y_1, y_2]$

$$\{p\}_m = \begin{Bmatrix} Z_{im} \\ M_{im} \\ Z_{jm} \\ M_{jm} \end{Bmatrix} = \begin{Bmatrix} \bar{x} - \frac{\bar{x}^3}{b^2} + \frac{\bar{x}^4}{2b^3} \\ \frac{\bar{x}^2}{2} - \frac{2\bar{x}^3}{3b} + \frac{\bar{x}^4}{4b^2} \\ \frac{\bar{x}^3}{b^2} - \frac{\bar{x}^4}{2b^3} \\ -\frac{\bar{x}^3}{3b} + \frac{\bar{x}^4}{4b^2} \end{Bmatrix} Q_o C_m$$

where

$$\bar{x}^n = x_2^n - x_1^n$$

$$C_m = \frac{1}{k_m} (\cos k_m y_1 - \cos k_m y_2)$$

NODAL
LINES

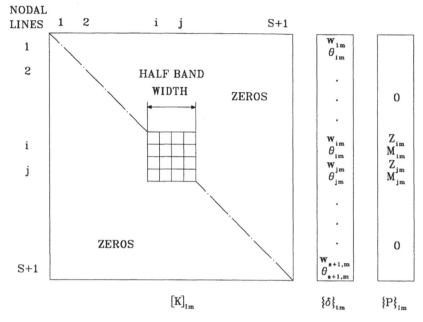

Figure 4. Expanded strip matrices.

a $N \times 1$ vector $\{P\}_{Im}$, respectively. These are constructed by inserting the related stiffness coefficients and loads in their proper locations and filling the remaining locations with zeros as shown in Figure 4. Thus, Equation 18 and Equation 19 can be written as

$$U_I = \frac{1}{2} \sum_{m=1}^{r} \{\delta\}_{tm}^T [K]_{Im} \{\delta\}_{tm} \tag{24}$$

$$W_I = - \sum_{m=1}^{r} \{\delta\}_{tm}^T \{P\}_{Im} \tag{25}$$

and Equation 22 and Equation 23 become

$$U_t = \frac{1}{2} \sum_{m=1}^{r} \{\delta\}_{tm}^T (\sum_{I=1}^{S} [K]_{Im}) \{\delta\}_{tm} = \frac{1}{2} \sum_{m=1}^{r} \{\delta\}_{tm}^T [K]_{tm} \{\delta\}_{tm} \tag{26}$$

$$W_t = - \sum_{m=1}^{r} \{\delta\}_{tm}^T (\sum_{I=1}^{S} \{P\}_{Im}) = - \sum_{m=1}^{r} \{\delta\}_{tm}^T \{P\}_{tm} \tag{27}$$

where $[K]_{tm}$ and $\{P\}_{tm}$ are the overall stiffness matrix and load vector of the entire structure respectively.

Since the stiffness matrix of every strip is symmetrical about its diagonal, the overall stiffness matrix is also symmetrical. Moreover, since the strain energy can not be negative, the overall stiffness matrix is positive-definite. Furthermore, it can be seen that the overall stiffness matrix has non-zero items in a narrow band symmetrical about the diagonal (for a simply supported plate, the half-bandwidth is four). These properties of the overall stiffness matrix enhance the efficiency of analysis and reduce the requirement for computer storage significantly.

The total potential energy of the entire plate is now expressed as

$$\Pi_t = U_t + W_t = \sum_{m=1}^{r} \left(\frac{1}{2}\{\delta\}_{tm}^T [K]_{tm}\{\delta\}_{tm} - \{\delta\}_{tm}^T \{P\}_{tm} \right) \qquad (28)$$

5.2.4. Minimization of Total Potential Energy

In order to obtain the best approximation, the values of the displacement parameters should be so chosen as to make the total potential energy of the plate become minimum. Therefore, the first derivative of this energy with respect to every displacement parameter must be zero, i.e.

$$\frac{\partial \Pi_t}{\partial w_{im}} = 0$$
$$\frac{\partial \Pi_t}{\partial \theta_{im}} = 0; \quad i = 1, N' \qquad (29)$$

where N' is the total number of nodal lines ($N' = S + 1$ for a plate). In a matrix form, Equation 29 can be written as

$$\frac{\partial \Pi_t}{\partial \{\delta\}_{tm}} = \{0\} \qquad (30)$$

Substituting Equation 28 into Equation 30 yields a set of linear algebraic equations for the m-th series term:

$$[K]_{tm}\{\delta\}_{tm} = \{P\}_{tm} \qquad (31)$$

which can be solved by means of standard subprogram for solution of linear equations with positive-definite band symmetric matrix.

After solving Equation 31 for the unknown displacement parameters $\{\delta\}_{tm}$ of all the series terms, the displacements and bending moments at any point inside the structure can be obtained from Equation 4 and Equation 12.

5.2.5. General Cases of End Support Conditions

In general case of end support conditions, the displacement function of a plate strip may take the following form:

$$w(x, y) = \sum_{m=1}^{r} [N] Y_m(y)\{\delta\}_m \tag{32}$$

where $[N]$ is the matrix of shape functions of x (see Equation 2), and $Y_m(y)$ is the m-th term of a series of functions of y which must satisfy the end support conditions a priori.

The most commonly used is the series of beam eigenfunctions which are derived from the solution of the beam vibration differential equation

$$\frac{d^4 Y}{dy^4} = \mu^4 Y \tag{33}$$

The general form of the beam eigenfunctions is

$$Y(y) = c_1 \sin(\mu y) + c_2 \cos(\mu y) + c_3 \sinh(\mu y) + c_4 \cosh(\mu y) \tag{34}$$

with the coefficients c_i to be determined by the end conditions. These have been worked out explicitly for the various end conditions and are listed below:

(a) Both ends simply supported, i.e., $Y(0) = Y''(0) = 0$ and $Y(l) = Y''(l) = 0$:

$$Y_m(y) = \sin(\mu_m y), \tag{35}$$

where

$$\mu_m = \frac{m\pi}{l}$$

(b) Both ends clamped, i.e., $Y(0) = Y'(0) = 0$ and $Y(l) = Y'(l) = 0$:

$$Y_m(y) = \sin(\mu_m y) - \sinh(\mu_m y) - \alpha_m[\cos(\mu_m y) - \cosh(\mu_m y)] \tag{36}$$

where

$$\alpha_m = \frac{\sin(\mu_m l) - \sinh(\mu_m l)}{\cos(\mu_m l) - \cosh(\mu_m l)}$$

Table 3. μ_1 to μ_{12} for clamped ends

$\mu_1 l$	4.730040744862704	$\mu_7 l$	23.56194490204045
$\mu_2 l$	7.853204624095837	$\mu_8 l$	26.70353755550818
$\mu_3 l$	10.99560783800167	$\mu_9 l$	29.84513020910325
$\mu_4 l$	14.13716549125746	$\mu_{10} l$	32.98672286269282
$\mu_5 l$	17.27875965739948	$\mu_{11} l$	36.12831551628262
$\mu_6 l$	20.42035224562606	$\mu_{12} l$	39.26990816987241

$\mu_m l$ are the solutions of equation $1 - \cos(\mu l)\cosh(\mu l) = 0$, the first twelve values are listed in Table 3. For m greater than 12, $\mu_m = (m + 0.5)\pi/l$ can be taken as a very close approximation.

For other types of support conditions, the reader may consult[1,2].

It can be proven that the beam eigenfunctions possess the valuable properties of orthogonality, i.e.

$$\int_0^l Y_m Y_n dy = 0 \qquad \text{for } m \neq n$$

$$\int_0^l Y_m'' Y_n'' dy = 0 \qquad \text{for } m \neq n$$

Utilization of these properties will result in a significant saving in computation effort for the calculation of stiffness matrices.

From the assumed displacement function in Equation 32, the curvature vector may be expressed in terms of the displacement parameters as:

$$\{\kappa\} = \left[\frac{\partial^2 w}{\partial x^2}, \frac{\partial^2 w}{\partial y^2}, -2\frac{\partial^2 w}{\partial x \partial y} \right]^T = \sum_{m=1}^r [B]_m \{\delta\}_m \qquad (37)$$

where

$$[B]_m = \begin{bmatrix} N_1'' Y_m(y) & N_2'' Y_m(y) & N_3'' Y_m(y) & N_4'' Y_m(y) \\ N_1 Y_m''(y) & N_2 Y_m''(y) & N_3 Y_m''(y) & N_4 Y_m''(y) \\ -2N_1' Y_m'(y) & -2N_2' Y_m'(y) & -2N_3' Y_m'(y) & -2N_4' Y_m'(y) \end{bmatrix} \qquad (38)$$

The moment vector, the strain energy of the strip and the potential energy of external loading can then be expressed as follows:

$$\{M\} = \sum_{m=1}^r [D][B]_m \{\delta\}_m \qquad (39)$$

$$U = \frac{1}{2} \sum_{m=1}^{r} \sum_{n=1}^{r} \{\delta\}_m^T \int_0^l \int_0^b [B]_m^T [D][B]_n dx dy \{\delta\}_n \tag{40}$$

$$W = - \sum_{m=1}^{r} \{\delta\}_m^T \int_0^l \int_0^b [N]^T q(x, y) Y_m(y) dx dy \tag{41}$$

After integration, Equation 40 and 41 become

$$U = \frac{1}{2} \sum_{m=1}^{r} \sum_{n=1}^{r} \{\delta\}_m^T [k]_{mn} \{\delta\}_n \tag{42}$$

$$W = - \sum_{m=1}^{r} \{\delta\}_m^T \{p\}_m \tag{43}$$

in which

$$[k]_{mn} = \int_0^l \int_0^b [B]_m^T [D][B]_n dx dy \tag{44}$$

and

$$\{p\}_m = \int_0^l \int_0^b [N]^T q(x, y) Y_m(y) dx dy \tag{45}$$

From Equation 38 it can be seen that Equation 44 includes the following integrals:

$$I_1 = \int_0^l Y_m Y_n dy \qquad I_2 = \int_0^l Y_m'' Y_n'' dy$$

$$I_3 = \int_0^l Y_m' Y_n' dy \qquad I_4 = \int_0^l Y_m Y_n'' dy$$

It is already known that integral I_1 and I_2 are equal to zero for $m \neq n$, whereas, integral I_3 and I_4 will not in general vanish for $m \neq n$. Consequently, the different series terms are coupled, i.e. $[K]_{mn} \neq [0]$ for $m \neq n$, unless the plate is simply supported on two opposite edges.

The overall matrix equation for the entire structure and all the series terms takes the following form:

$$\begin{bmatrix} [K]_{t11} & [K]_{t12} & \cdots & [K]_{t1r} \\ [K]_{t21} & [K]_{t22} & \cdots & [K]_{t2r} \\ \cdot & \cdot & \cdots & \cdot \\ [K]_{tr1} & [K]_{tr2} & \cdots & [K]_{trr} \end{bmatrix} \begin{Bmatrix} \{\delta\}_{t1} \\ \{\delta\}_{t2} \\ \cdot \\ \{\delta\}_{tr} \end{Bmatrix} = \begin{Bmatrix} \{P\}_{t1} \\ \{P\}_{t2} \\ \cdot \\ \{P\}_{tr} \end{Bmatrix} \tag{46}$$

In programming, in order to take advantage of banded nature of the stiffness matrix, the displacement parameters should be grouped in the following sequence:

$$\{\delta\}_t = [..., w_{i1}, w_{i2}, ..., w_{ir}, \theta_{i1}, \theta_{i2}, ..., \theta_{ir}, ...]^T \tag{47}$$

Once Equation 46 is solved for all the unknown displacement parameters (related to all nodal lines and all series terms), the displacements and moments at any point of the structure can readily be calculated using Equation 32 and Equation 39.

5.3. FLAT SHELL STRIP

5.3.1. Displacement Functions and Strip Matrices

A prismatic folded plate structure or box girder bridge (Figure 5) is an assembly of rectangular plates that can offer resistance to both bending and in-plane loadings. This type of structure can be analyzed using the flat shell strip that is capable of simulating both the bending and in-plane deformations. The displacement parameters of the strip are chosen as

$$\{\delta\}_m = [u_{im}, v_{im}, w_{im}, \theta_{im}, u_{jm}, v_{jm}, w_{jm}, \theta_{jm}]^T \tag{48}$$

in which u and v represent the in-plane displacements in the x and y directions, respectively (Figure 6).

If the both ends of structure are simply supported, i.e. the end supports are rigid in the transverse directions and totally flexible in the longitudinal direction as shown below

$$u = 0, \quad w = 0, \quad \sigma_y = 0 \quad \text{and} \quad M_y = 0 \quad \text{at} \quad y = 0 \quad \text{and} \quad y = l$$

the displacements in the middle plane of the strip can be interpolated as follows:

$$u = \sum_{m=1}^{r}((1 - X)u_{im} + Xu_{jm}) \sin k_m y$$

$$v = \sum_{m=1}^{r}((1 - X)v_{im} + Xv_{jm}) \cos k_m y \tag{50}$$

$$w = \sum_{m=1}^{r}(N_1(x)w_{im} + N_2(x)\theta_{im} + N_3(x)w_{jm} + N_4(x)\theta_{jm}) \sin k_m y$$

NODAL LINE   FINITE STRIP

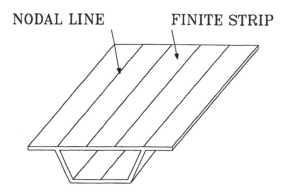

Figure 5. Box structure and finite strip.

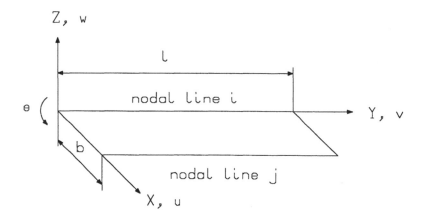

Figure 6. Flat shell strip.

where $X = x/b$, $k_m = m\pi/l$, $N_1(x)$ to $N_4(x)$ are Hermitian polynomials defined in Equation 2, l and b are respectively the length and width of the strip; and r is the total number of series terms used in analysis.

Above equation can also be written in a matrix form as

$$\{f\} = [u, v, w]^T = \sum_{m=1}^{r} [N]_m \{\delta\}_m \tag{51}$$

Including both bending and in-plane actions, the following strain-displacement relationship and stress-strain relationship are applicable[6]:

$$\epsilon_x = \frac{\partial u}{\partial x}$$

$$\epsilon_y = \frac{\partial v}{\partial y}$$

$$\gamma_{xy} = \frac{\partial u}{\partial y} + \frac{\partial v}{\partial x}$$

$$\chi_x = \frac{\partial^2 w}{\partial x^2}$$

$$\chi_y = \frac{\partial^2 w}{\partial y^2}$$

$$\chi_{xy} = -2\frac{\partial^2 w}{\partial x \partial y}$$

(52)

$$\{\sigma\} = \begin{Bmatrix} N_x \\ N_y \\ N_{xy} \\ M_x \\ M_y \\ M_{xy} \end{Bmatrix} = \begin{bmatrix} K_x & K_1 & & & & \\ K_1 & K_y & & & & \\ & & K_{xy} & & & \\ & & & D_x & D_1 & \\ & & & D_1 & D_y & \\ & & & & & D_{xy} \end{bmatrix} \begin{Bmatrix} \epsilon_x \\ \epsilon_y \\ \gamma_{xy} \\ \chi_x \\ \chi_y \\ \chi_{xy} \end{Bmatrix} = [D]\{\epsilon\}$$

(53)

where

$$K_x = E_x h/(1 - \nu_x \nu_y) \qquad K_y = E_y h/(1 - \nu_x \nu_y)$$
$$K_{xy} = E_{xy} h \qquad\qquad K_1 = \nu_x K_x = \nu_y K_y$$
$$D_x = E_x h^3/12(1 - \nu_x \nu_y) \qquad D_y = E_y h^3/12(1 - \nu_x \nu_y)$$
$$D_{xy} = E_{xy} h^3/12 \qquad\qquad D_1 = \nu_x D_x = \nu_y D_y \qquad (54)$$

where h is the thickness of the plate.

By substituting Equation 50 into Equation 52, the strains can be expressed in terms of displacement parameters in the form

$$\{\epsilon\} = [\epsilon_x, \epsilon_y, \gamma_{xy}, \chi_x, \chi_y, \chi_{xy}]^T = \sum_{m=1}^{r} [B]_m \{\delta\}_m \qquad (55)$$

where $[B]_m$ is strain matrix.

By applying the principle of minimum total potential energy and following the procedure described in the previous section, the stiffness matrix and the load vector of the strip can be obtained as

$$[k]_m = \int_0^l \int_0^b [B]_m^T [D][B]_m \, dx \, dy \qquad (56)$$

and

$$\{p\}_m = \int_0^l \int_0^b [N]_m^T [p_x, p_y, q]^T \, dx \, dy \qquad (57)$$

where p_x, p_y and q are the distributed in-plane and transverse loads, respectively.

5.3.2. Coordinate Transformation

The above derivations are carried out in a local coordinate system, wherein the x and y axes coincide with the mid-surface of a strip. In folded plate structures, any two plates will in general meet at an angle, and in order to assemble the stiffness matrices, the displacement vectors and the load vectors of non-coplanar strips, a global coordinate system common to all the strips must be established.

In Figure 7 x, y and z are individual coordinates of a strip, and $\bar{x}$, $\bar{y}$ and $\bar{z}$ are the global coordinates. y and $\bar{y}$ coincide with the intersection line of two adjoining strips. The local and global displacement components are related as

$$u_{im} = \bar{u}_{im} \cos \beta + \bar{w}_{im} \sin \beta$$

$$v_{im} = \bar{v}_{im}$$

$$w_{im} = -\bar{u}_{im} \sin \beta + \bar{w}_{im} \cos \beta \qquad (58)$$

$$\theta_{im} = \bar{\theta}_{im}$$

This relationship is also applicable for nodal line j. Thus, for a strip, the following transformation is obtained:

$$\{\delta\}_m = \begin{bmatrix} [t] & 0 \\ 0 & [t] \end{bmatrix} \{\bar{\delta}\}_m = [T]\{\bar{\delta}\}_m \qquad (59)$$

where

$$[t] = \begin{bmatrix} \cos \beta & 0 & \sin \beta & 0 \\ 0 & 1 & 0 & 0 \\ -\sin \beta & 0 & \cos \beta & 0 \\ 0 & 0 & 0 & 1 \end{bmatrix} \qquad (60)$$

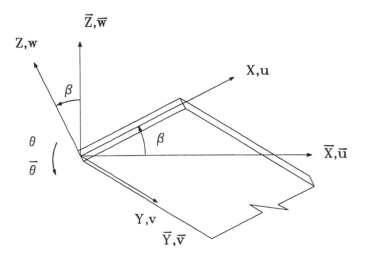

Figure 7. Coordinate transformation.

Then, the total potential energy of the strip can be expressed in terms of the global displacement vector and global load vector as follows

$$
\begin{aligned}
\Pi &= \frac{1}{2} \sum_{m=1}^{r} \{\delta\}_m^T [k]_m \{\delta\}_m - \sum_{m=1}^{r} \{\delta\}_m^T \{p\}_m \\
&= \frac{1}{2} \sum_{m=1}^{r} \{\bar\delta\}_m^T [T]^T [k]_m [T] \{\bar\delta\}_m - \sum_{m=1}^{r} \{\bar\delta\}_m^T [T]^T \{p\}_m \quad (61) \\
&= \frac{1}{2} \sum_{m=1}^{r} \{\bar\delta\}_m^T [\bar k]_m \{\bar\delta\}_m - \sum_{m=1}^{r} \{\bar\delta\}_m^T \{\bar p\}_m
\end{aligned}
$$

from which the stiffness matrix and the load vector of the strip in the global system are obtained as:

$$
[\bar k_m] = [T]^T [k_m][T] \quad (62)
$$

$$
\{\bar p\}_m = [T]^T \{p\}_m \quad (63)
$$

By following the assembling and minimization procedures described earlier, a set of linear algebraic equations for the entire box structure is established as

$$
[\bar K]_{tm} \{\bar\delta\}_{tm} = \{\bar P\}_{tm} \quad (64)
$$

After the global displacement parameters related to all the nodal lines and all the series terms are obtained from Equation 64, Equation 59 is used to transform back to the local coordinate system. The displacements, internal forces and moments at any point in the structure can then be computed in the local coordinate system of each strip.

5.4. COMBINED FS/FE ANALYSIS OF IRREGULAR PLATES

In this analysis, the regular region of a rectangular plate is modelled by finite strips presented in Section 5.2, while the irregular region is divided into a number of finite elements, and the both regions are connected by a row of transition elements.

5.4.1. Finite Elements for Irregular Region

The conforming rectangular plate elements (Figure 8)[7] are employed for the present analysis. Each element has four nodes, each of which has four degrees of freedom as below:

$$\{\delta\} = \left(w, \frac{\partial w}{\partial x}, \frac{\partial w}{\partial y}, \frac{\partial^2 w}{\partial x \partial y} \right)^T$$

The deflection within the element is written as:

$$
\begin{aligned}
w = {} & f_1(\xi) f_1(\eta) w_i + b g_1(\xi) f_1(\eta) \frac{\partial w_i}{\partial x} + a f_1(\xi) g_1(\eta) \frac{\partial w_i}{\partial y} \\
& + a b g_1(\xi) g_1(\eta) \frac{\partial^2 w_i}{\partial x \partial y} + f_1(\xi) f_2(\eta) w_j + b g_1(\xi) f_2(\eta) \frac{\partial w_j}{\partial x} \\
& + a f_1(\xi) g_2(\eta) \frac{\partial w_j}{\partial y} + a b g_1(\xi) g_2(\eta) \frac{\partial^2 w_j}{\partial x \partial y} + f_2(\xi) f_1(\eta) w_k \\
& + b g_2(\xi) f_1(\eta) \frac{\partial w_k}{\partial x} + a f_2(\xi) g_1(\eta) \frac{\partial w_k}{\partial y} + a b g_2(\xi) g_1(\eta) \frac{\partial^2 w_k}{\partial x \partial y} \\
& + f_2(\xi) f_2(\eta) w_l + b g_2(\xi) f_2(\eta) \frac{\partial w_l}{\partial x} + a f_2(\xi) g_2(\eta) \frac{\partial w_l}{\partial y} \\
& + a b g_2(\xi) g_2(\eta) \frac{\partial^2 w_l}{\partial x \partial y}
\end{aligned}
\tag{65}
$$

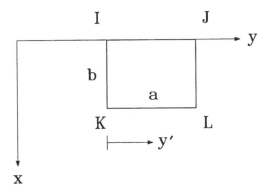

Figure 8. Rectangular finite element.

where $\xi = x/b$ and $\eta = y'/a$; $f_1(\xi)$, $f_2(\xi)$, $g_1(\xi)$ and $g_2(\xi)$ are Hermitian polynomials, which have already been introduced in Equation 2, but here written in a different pattern:

$$f_1(\xi) = 1 - 3\xi^2 + 2\xi^3, \quad f_2(\xi) = 3\xi^2 - 2\xi^3,$$
$$g_1(\xi) = \xi - 2\xi^2 + \xi^3, \quad g_2(\xi) = \xi^3 - \xi^2, \quad (\xi = x/b) \tag{66}$$

This type of element meets the continuity requirement between any adjacent elements not only for the deflection w and the tangential derivative $\frac{\partial w}{\partial s}$ but also for the normal derivative $\frac{\partial w}{\partial n}$. Therefore, this element is a conforming plate bending element, and it achieves considerable improvement in accuracy in comparison with rectangular elements having only 3 degrees of freedom at each node.

5.4.2. Transition Element

A row of rectangular transition elements is used to connect the finite element region with the finite strip region. One side of each transition element coincides with the nodal line of adjacent strip and has the same degrees of freedom. On the opposite side there are two corner nodes which are attached to the nodes of adjacent finite elements and have the same degrees of freedom as these finite element nodes (Figure 9). Inside the transition element, the deflection w is expressed in terms of degrees of freedom of nodal line and

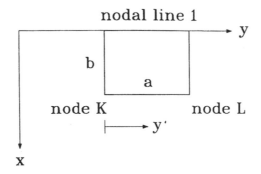

Figure 9. Transition element.

two nodes by their corresponding shape functions in the form:

$$w = \sum_{m=1}^{r}(f_1(\xi)w_{1m} + bg_1(\xi)\theta_{1m})Y_m(y) + f_2(\xi)f_1(\eta)w_k$$

$$+ bg_2(\xi)f_1(\eta)\frac{\partial w_k}{\partial x} + af_2(\xi)g_1(\eta)\frac{\partial w_k}{\partial y} + abg_2(\xi)g_1(\eta)\frac{\partial^2 w_k}{\partial x \partial y}$$

$$+ f_2(\xi)f_2(\eta)w_l + bg_2(\xi)f_2(\eta)\frac{\partial w_l}{\partial x} + af_2(\xi)g_2(\eta)\frac{\partial w_l}{\partial y}$$

$$+ abg_2(\xi)g_2(\eta)\frac{\partial^2 w_l}{\partial x \partial y} \tag{67}$$

where $\xi = x/b$ and $\eta = y'/a$.

It should be noted that the nodal line 1 and nodes k, l have the same type of shape function in the direction x, which are Hermitian cubic polynomials, in order to simulate some basic deformation patterns more accurately, such as one dimensional bending in the direction y, torsion about the axis y, etc..

The stiffness matrix and the load vector can be obtained in accordance with standard finite element formulation. Simply assembling the stiffness matrices and load arrays of these transition elements with those of finite strips and finite elements will lead immediately to the formulation for the whole plate structure.

5.4.3. Numerical Examples

5.4.3.1. *Clamped square plate under uniform load*

This example is chosen to verify the present method. The square plate is clamped along four edges and is subjected to uniform load. The length of

Table 4. Deflection and moments in clamped square plate

No.of Strips	2	3	4	
No.of Sym.Terms	3	3	5	Theory
Mesh of elements	2 by 4	3 by 6	4 by 8	
w_{max}	0.0113105	0.0111349	0.0110848	0.0110074
M_x at A	−4.390	−4.747	−4.896	
M_x at B	−4.719	−4.916	−5.001	
M_y at C	−5.111	−4.915	−5.071	−5.130
M_x at E	2.496	2.380	2.331	
M_y at E	2.445	2.376	2.322	2.375

each side is 10.0 m, the thickness is 0.5 m. The material properties are $E = 100000 MPa$ and $v = 0.3$. The intensity of the load is $q = 1.0 MN/m^2$. Half the plate is divided into finite strips, the other half into finite elements. The mesh is shown in Figure 10 and the results are listed in Table 4. The analytical results are also given in this table for comparison. From this table it can been seen that the method achieves satisfactory accuracy in evaluating maximum deflection and bending moment.

5.4.3.2. Square plate supported by walls and columns

A square plate with a rectangular opening is supported by walls and columns (Figure 11). The length of each side is 9.0 m and the thickness is 0.2 m. The material properties are $E = 25000$ MPa and $v = 0.15$. The plate is subjected to uniform load of intensity $q = 10$ kN/m^2.

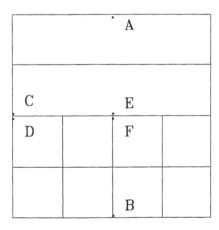

Figure 10. Square plate.

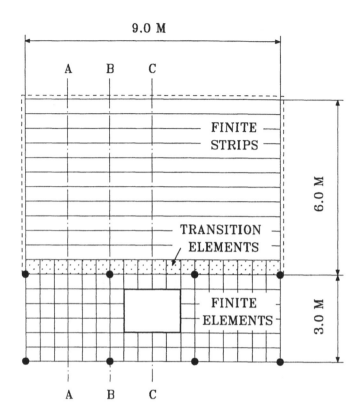

Figure 11. Plate supported by walls and columns.

The part supported by columns is divided into 6 by 18 conforming rectangular plate elements with 4 degrees of freedom at each node; the rest of the plate is divided into 11 strips and one row of transition elements that connect the strips to the finite elements. Five symmetrical series terms are used. For simplicity, the walls are regarded as simple supports, and the columns as point supports.

The resulting deflection along line C-C and the bending moment M_x along the lines B-B and C-C are shown in Figure 12. The CPU time spent for the analysis is 4.2 sec. on a Mainframe AMDAHL-5860.

The structure was also analyzed using an 18 × 18 array of the above-mentioned elements at a cost of 12.1 sec. of CPU time. However, the results show no discernible difference from the previous solution when the results of both methods are drawn on the same figure.

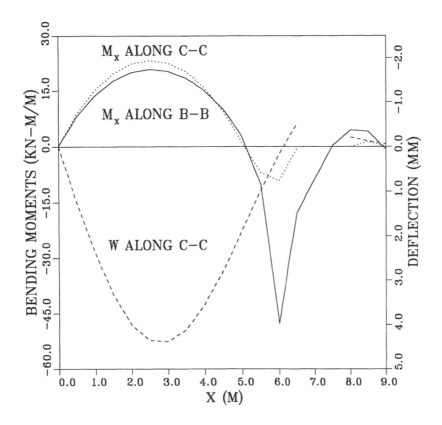

Figure 12. Deflection and bending moments of plate in Figure 11.

5.5. COMBINED FS/BE ANALYSIS OF IRREGULAR PLATES

5.5.1. Boundary Elements for Irregular Region

In this approach, the boundary of the irregular region of plate is divided into a number of boundary elements. Each element has between 2 and 11 nodes, and all the boundary functions along the element are expressed in terms of their nodal values by Lagrange interpolation.[8] A source point is located at a small distance from the boundary on the outside normal line passing through each node. Applying a unit force and a unit normal moment at each source point respectively and implementing the Maxwell-Betti theorem provide the two integration equations:

$$\int_s (WV^f + \Theta M^f)\, ds + \sum_{i=1}^{N_c} W_i C_i^f = \int_A pw^f\, dA$$
$$+ \int_s (W^f V + \Theta^f M)\, ds + \sum_{i=1}^{N_c} W_i^f C_i \qquad (68)$$

$$\int_s (WV^m + \Theta M^m)\, ds + \sum_{i=1}^{N_c} W_i C_i^m = \int_A pw^m\, dA$$
$$+ \int_s (W^m V + \Theta^m M)\, ds + \sum_{i=1}^{N_c} W_i^m C_i \qquad (69)$$

where
   A and s denote the plate area and the boundary coordinate,
   w is deflection in domain,
   W and Θ are boundary deflection and normal rotation,
   M, V and C are boundary moment, equivalent shear and corner force,
   f and m are superscripts identifying the fundamental solutions for the unit force and the unit moment respectively,
   N_c is the number of corners.
   All four boundary functions W, Θ, M and V are unknowns on the interface between the transition strip and the neighboring boundary element, but among them only two are unknown on the other boundaries after imposing displacement and force boundary conditions.
   At the ends of interface, double nodes are used[9] (Figure 13). Both nodes A and B have the same coordinates, but may have different boundary conditions. In order to make the number of unknowns at nodes A and B match the number of the boundary integration equations, the corner forces at these two nodes may be expressed in terms of the normal rotations of their respective element in the form:

$$C_A = D(1 - v)\left(\frac{\partial^2 w}{\partial s \partial n}\right)^{(1)} = D(1 - v)\sum_{j=1}^{N_1} \frac{\partial L_j^{(1)}}{\partial s}\Theta_j \qquad (70)$$

$$C_B = -D(1 - v)\left(\frac{\partial^2 w}{\partial s \partial n}\right)^{(2)} = -D(1 - v)\sum_{j=1}^{N_2} \frac{\partial L_j^{(2)}}{\partial s}\Theta_j \qquad (71)$$

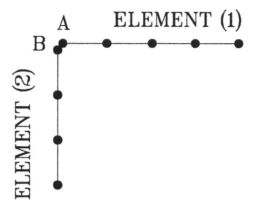

Figure 13. Double nodes.

where (1) and (2) are the superscripts identifying the elements ending at node A and starting at node B respectively; N_1 and N_2 are the numbers of nodes on boundary elements 1 and 2 respectively; L_j is the Lagrange shape function of node j and is expressed as:

$$L_j(s) = \prod_{\substack{i = 1 \\ i \neq j}}^{N} \frac{s - s_i}{s_j - s_i} \qquad (72)$$

Substituting the expressions for fundamental solutions into the boundary integral equations and implementing the boundary element discretization technique[8] lead to the following boundary element matrices:

$$[H^1, H_I^1] \begin{Bmatrix} U^1 \\ U_I^1 \end{Bmatrix} = [G^1, G_I^1] \begin{Bmatrix} P^1 \\ P_I^1 \end{Bmatrix} + \{B^1\} \qquad (73)$$

where
   U denotes the values of deflection and normal rotation at all the nodes,
   P represents the values of equivalent shear and normal moment at all the nodes,
   B is the effect of the loads in domain,
   1 is the superscript identifying the boundary element region,
   I is the subscript defining the interface between the transition strip and the boundary elements.

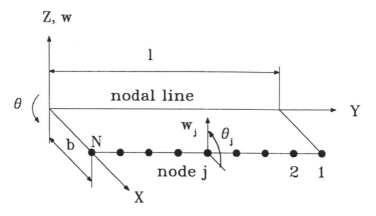

Figure 14. Transition strip of plate.

For each node on the interface there are 4 unknowns (W, Θ, V and M)́ versus 2 equations. Therefore, two more equations are required for each of these nodes.

5.5.2. Transition Strip and Combined Solution

A transition strip is inserted between the finite strip region and the boundary element region and connects the two regions together. One side of the transition strip coincides with the nodal line of the adjacent strip and has the same degrees of freedom as this nodal line. On the opposite side there are a number of nodes which are attached to the nodes of the neighboring boundary elements and have the same displacement parameters (deflection W and normal rotation Θ) (Figure 14). Inside the transition strip the deflection w is expressed in terms of the degrees of freedom of the nodal line and the nodes by their corresponding shape functions in the form:

$$w = \sum_{m=1}^{r}(f_1(\xi)w_{1m} + bg_1(\xi)\theta_{1m})Y_m(y) + \sum_{j=1}^{N}L_j(f_2(\xi)w_j + ag_2(\xi)\theta_j)$$

(74)

where

N is the number of nodes of this strip along the interface,

L_j is the Lagrange shape function of node j.

It should be noted that the nodal line and nodes 1 to N have the same type of shape functions in the direction x, which are Hermitian cubic polynomials, in order to simulate some basic deformation patterns more accurately as noted earlier.

Using the principle of minimum total potential energy, the stiffness matrix and the load vector of the transition strip are generated.

Assembling the stiffness matrix and load vector of the transition strip with those of the finite strips and applying static condensation technique[10] will transform the finite strips and transition strip into a substructure having only the degrees of freedom of the nodes on the interface. The matrix equation of this substructure is of the form:

$$[K]\{U_I^2\} = \{F_I^2\} + \{B^2\} \tag{75}$$

where

F_I^2 is the vector of unknown interaction nodal forces,

B represents the effects of external load,

2 is the superscript identifying the finite strip and transition strip region.

At all nodes on the interface, the displacement on both sides must be the same, and the interaction force must be the same in magnitude and opposite in direction. These requirements yield the following relationships

$$\{U_I^2\} = \{U_I^1\} \tag{76}$$

$$\{F_I^2\} = -[M]\{P_I^1\} \tag{77}$$

where $[M]$ is the transformation matrix from the distributed boundary forces on the interface boundary element to the nodal forces at the nodes of the transition strip, and its elements M_{ij} are given by

$$M_{ij} = \int_e L_i(y) L_j(y) \, dy$$

Substituting Equation 76 and Equation 77 into Equation 75 gives a set of equations:

$$[K]\{U_I^1\} = -[M]\{P_I^1\} + \{B^2\} \tag{78}$$

Combining matrix Equations 73 and 78 together leads to a set of equations that are just sufficient in number to solve for all the unknowns in the boundary element region, including the unknowns at the nodes on the interface; these latter can be substituted back into the matrix of the finite strip and transition strip region, and all the unknown nodal parameters in this region can then be readily obtained. Thereafter it is a straightforward procedure to compute the displacement and stress components at any points of interest.

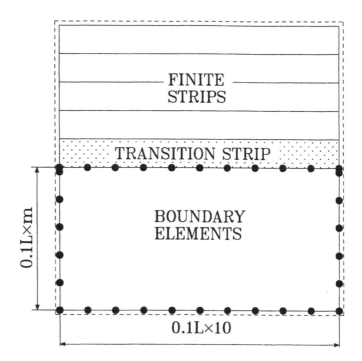

Figure 15. Simply supported square plate under uniform load.

5.5.3. Numerical Examples

5.5.3.1. Simply supported square plate under uniform load

This example is chosen to verify the present method. A square plate is hinged along four edges and subjected to uniform load. The length of each side is L, Poisson's ratio is $\nu = 0.3$, Flexural rigidity is D, and the intensity of the load is q. Three different meshes are used to analyze the plate (Figure 15). The upper part of each mesh is divided into finite strips with equal width 0.1L. Five symmetrical terms are used in the series of longitudinal shape functions. The boundary of the lower part is simulated by 4 boundary elements with equal nodal spacing 0.1L. Both parts are connected by a transition strip with a width of 0.1L. The numerical results are listed in Table 5. In this table, W_{max} and $(M_x)_{max}$ are the deflection and bending moment at the plate center respectively, V_{max} is the equivalent shear force at the center of the bottom edge, and R is the corner force at the lower corners. In comparison with plate theory[6], it can be seen that the accuracy of the present method is satisfactory.

Table 5. Deflection and stresses in simply supported square plate

Mesh	w_{max} $= \alpha q L^4 / D$ α	$(M_x)_{max}$ $= \beta q L^2$ β	V_{max} $= \delta q L$ δ	R $n q L^2$ n
m=3	0.004063	0.04804	0.4205	0.0618
m=5	0.004062	0.04790	0.4205	0.0646
m=7	0.004063	0.04789	0.4205	0.0654
Theory	0.00406	0.0479	0.420	0.065

5.5.3.2. *Simply supported square plate with an opening and a skew corner*

A square plate is simply supported along all four sides and subjected to uniform load $q = 10$ KN/m^2. The length of each side is 10.0 m, and the thickness 0.2 m. The material properties arc $E = 25000$ Mpa and $v = 0.15$. One corner is cut off at 45 degrees, and a square opening is located near another corner, as shown in Figure 16.

The part including the opening and the skew corner is analyzed by 9 boundary elements with 50 nodes whilst the rest of the plate is analyzed by 5 finite strips and one transition strip with 10 series terms.

The resulting bending moments M_x and M_y along line A-B-C are depicted in Figure 17.

The plate was also analyzed by the boundary element method alone, with 9 elements and 58 nodes. The results of this analysis if plotted on Figure 17 would be indistinguishable from those already shown.

5.6. COMBINED BE/FS ANALYSIS OF SLAB GIRDER BRIDGES

As mentioned earlier, if a slab girder or box girder bridge is subjected to moving wheel loads and the local bending moments are concerned, the slab can be analyzed by the boundary element method and the girders are modelled by flat shell strips, which are connected to the slab by transition strips (Figure 18).

5.6.1. Boundary Elements for Slab Panel

In this approach, the boundary of each slab panel is divided into a number of boundary elements which include actions of both in-plane stress and plate bending. Each element has 2 to 11 nodes, and all the boundary functions

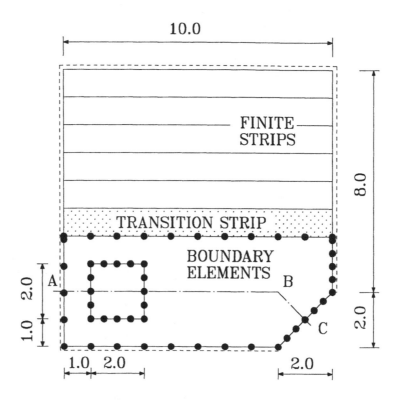

Figure 16. Plate with opening and skew corner.

along the element are expressed in terms of their nodal values by Lagrange interpolation as follows:

$$[u, v, w, \theta, P_x, P_y, V, M]^T = \sum_{j=1}^{N} L_j(s)[u_j, v_j, w_j, \theta_j, P_{xj}, P_{yj}, V_j, M_j]^T$$

(79)

where $L_j(s)$ is the Lagrange shape function given in Equation 72;

  N is the number of nodes in the element;

  s is the boundary coordinate of the element;

  u, v, w are the boundary displacements and θ is the normal rotation;

  P_x and P_y are the boundary in-plane forces;

  M is the boundary normal bending moment;

  V is the equivalent transverse shear. It also includes the effect of corner forces.

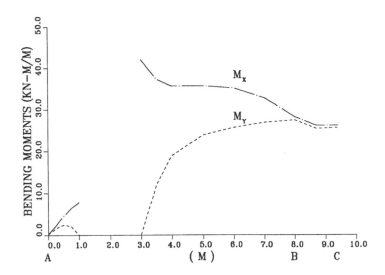

Figure 17. Bending moments along A-B-C.

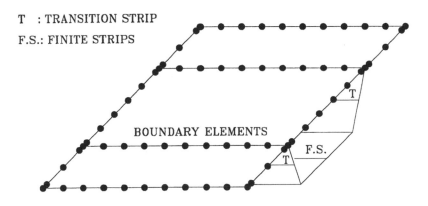

Figure 18. Combined FS/BE analysis of bridge.

At the ends of each girder, double nodes (Figure 13) are used: both have the same coordinates but different boundary conditions.

For the in-plane deformation, each boundary node is selected as a source point. For the plate bending, a different source point, located at a short distance from the boundary on the outside normal line passing through each node, is selected. Applying unit forces and unit moment at the source points of each node, and implementing the Maxwell-Betti theorem, provide four integration equations for every node:

$$c_x u + \int_s (u P_x^{f_x} + v P_y^{f_x}) ds = \int_s (u^{f_x} P_x + v^{f_x} P_y) ds$$

$$c_y v + \int_s (u P_x^{f_y} + v P_y^{f_y}) ds = \int_s (u^{f_y} P_x + v^{f_y} P_y) ds$$

$$\int_s (w V^{f_z} + \theta M^{f_z}) ds + \sum_{i=1}^{N_c} w_i C_i^{f_z} = \int_a p_z w^{f_z} dA + \int_s (w^{f_z} V + \theta^{f_z} M) ds$$

$$\int_s (w V^m + \theta M^m) ds + \sum_{i=1}^{N_c} w_i C_i^m = \int_a p_z w^m dA + \int_s (w^m V + \theta^m M) ds$$

$$(80)$$

where c_x and c_y are coefficients which may be determined by the conditions of the rigid body motion[9];

N_c is the number of corners;

C is the corner force of plate bending;

p_z is the vertical distributed load on the slab;

A denotes the area of plate midplane;

f_x, f_y, f_z and m are the superscripts identifying the fundamental solutions of the unit forces in x, y and z directions and the unit normal moment respectively.

By substituting Equation 79 and the expressions for the fundamental solutions into the boundary integration Equation 80 for all the boundary nodes, and performing the boundary and area integrations analytically[8], the following boundary element matrix equation is obtained for each panel:

$$[H^1, H_I^1] \left\{ \begin{matrix} U^1 \\ U_I^1 \end{matrix} \right\} = [G^1, G_I^1] \left\{ \begin{matrix} P^1 \\ P_I^1 \end{matrix} \right\} + \{B^1\} \qquad (81)$$

where

U denotes the values of the displacements and normal rotation at all the boundary nodes;

P represents the values of the boundary forces and normal moment at all the nodes;

B is the effect of the loads in the domain;

I is the subscript defining the interfaces between slab and girders;

1 is the superscript identifying the boundary element analysis.

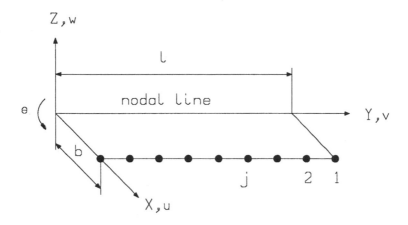

Figure 19. Transition strip of flat shell.

For each panel, Equation 81 includes four equations for each node. For every node beyond the interfaces, only four variables out of the following eight $(u_j, v_j, w_j, \theta_j, P_{xj}, P_{yj}, V_j$ and $M_j)$ are unknown after imposing the displacement and force boundary conditions. However, for every node at the interfaces, the above eight parameters are all unknown. Therefore, more equations are required for interface nodes. These equations can be obtained by taking into consideration the compatibility and equilibrium conditions across the interfaces as introduced in the following section.

5.6.2. Transition Strips and Combined Solution

Transition strips are inserted between the finite strips and the boundary elements to connect them together. One side of a transition strip coincides with the nodal line of the adjacent finite strip and has the same degrees of freedom of this nodal line. On the opposite side there are a number of nodes which are attached to the nodes of neighboring boundary elements and have the same displacement parameters (displacements u, v, w and normal rotation θ) as those of the nodes of the boundary elements (Figure 19). Within each transition strip, the displacements are expressed in terms of the degrees of freedom of the nodal line, and of the nodes. The corresponding shape functions are as follows:

$$u = \sum_{m=1}^{r}[(1-\xi)u_{1m}]Y_m(y) + \sum_{j=1}^{N}\xi L_j(y)u_j$$

$$v = \sum_{m=1}^{r}[(1-\xi)v_{1m}]Y'_m(y)/\mu_m + \sum_{j=1}^{N}\xi L_j(y)v_j$$

$$w = \sum_{m=1}^{r}(f_1(\xi)w_{1m} + bg_1(\xi)\theta_{1m})Y_m(y)+$$
$$+ \sum_{j=1}^{N}L_j(f_2(\xi)w_j + bg_2(\xi)\theta_j)$$

$$(82)$$

The stiffness matrix of the transition strip can be obtained using the principle of the minimum total potential energy. After assembling the stiffness matrices of the transition strips with those of the finite strips, the combined finite strips and transition strips are transformed into a substructure by applying the static condensation technique[10] so that the resulting substructure only has degrees of freedom at the nodes on the interfaces. The matrix equation of this substructure has the following form:

$$[K]\{U_I^2\} = \{F_I^2\} + \{B^2\} \tag{83}$$

where

F_I^2 is the vector of the unknown interaction nodal forces;

B represents the effect of the external load on the substructure;

2 is the superscript identifying the finite strip and transition strip substructure.

In general, two panels and one girder meet on each interface as shown in Figure 18. At all the nodes on the interface, the displacements of the panels and the girder should be the same, and the interaction forces should satisfy the equilibrium conditions. This yields the following relationships:

$$\{U_I^2\} = \{U_I^{1l}\} = \{U_I^{1r}\} = \{U_I^1\} \tag{84}$$

and

$$\{F_I^2\} + [M](\{P_I^{1l}\} + \{P_I^{1r}\}) = \{0\} \tag{85}$$

where the superscripts $1l$ and $1r$ refer respectively to the left and the right panels of the interface, and $[M]$ is the transformation matrix from the distributed boundary forces on the interface boundary elements to the nodal forces at the nodes of the transition strips. The transformation coefficient between node i and node j is

$$M_{ij} = \int_e L_i(y) L_j(y) \, dy$$

Substituting Equation 84 and Equation 85 into Equation 83 gives a set of equations:

$$[K]\{U_I^1\} = -[M](\{P_I^{1l}\} + \{P_I^{1r}\}) + \{B^2\} \tag{86}$$

Considering all the panels and girders, and using the Equation 81 and Equation 86 together, will yield enough equations for solving all the unknowns of the boundary elements, including those at the nodes on the interfaces. By backsubstituting into the matrix of the finite strip and transition strip substructure, all the unknown nodal parameters are readily obtained. Then, the displacement and the stress components at any points of interest are computed in a straightforward procedure.

5.6.3. Numerical Examples

In order to illustrate the proposed methodology, a twin-girder rectangular bridge and a twin-girder skewed bridge are analyzed in the following examples, though the method is general enough to be applicable to multi-panel slab-on-girder or box-girder bridges.

5.6.3.1. Rectangular slab-on-girder bridge under a wheel load

This example is chosen to verify the present approach. A simply supported rectangular slab-on-girder bridge is shown in Figure 20. The material properties are $E = 25000$ MPa and $\nu = 0.15$. A vertical wheel load, $P = 11.25 \times 9.81$ kN, distributed uniformly on a small patch[11] is applied at the center of the slab as illustrated in Figure 20.

The bridge is first analyzed by the finite strip method. The entire slab is divided into 14 third-order flat shell strips, and each girder into two strips. Fifty symmetrical series terms are taken in the analysis. Little improvement can be obtained if more strips or series terms are employed.

Then, the bridge is analyzed by the present method with the mesh shown in Figure 21.

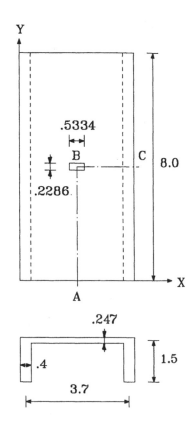

Figure 20. Rectangular slab on 2 girders (*m*).

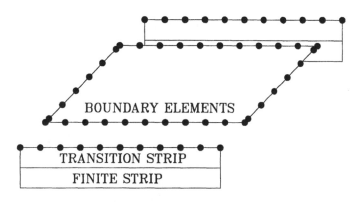

Figure 21. Mesh of combined analysis.

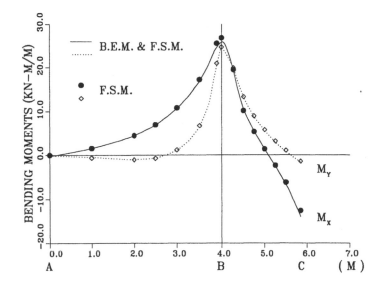

Figure 22. Bending moments along A-B-C.

The resulting bending moments along line A-B-C by both methods are depicted in Figure 22. The maximum bending moments occurring at the center of the wheel loading area are $M_x = 25.81$ kN-m/m and $M_y = 24.39$ kN-m/m by the present method comparatively to $M_x = 26.82$ kN-m/m and $M_y = 24.69$ kN-m/m by the finite strip analysis.

It can be seen that the agreement between the two methods is satisfactory. However, in order to achieve the same accuracy, the finite strip method needed 3800 degrees of freedom, while the present method employed only 224 degrees of freedom for its final solution.

If the two girders are replaced by two simply supported edges or two clamped edges, the resulting maximum bending moments are listed in Table 6. The results show that the replacements either overestimate or underestimate the maximum bending moments with errors of 10 to 20 percent. This suggests that the flexibility of the girder should not be neglected in the local effect analysis of the wheel load.

5.6.3.2. *Skewed slab-on-girder bridge under a wheel load*

In this example, a number of skewed slab-on-girder bridges is analyzed to investigate the effect of the skew angle on the local bending moments due to a wheel loading.

Table 6. Effect of simplification of girders

Simplification of Two Girders	M_x kN-m/m	M_y kN-m/m	M_{xy} kN-m/m
Simple Supports	30.79	27.11	0.00
Clamped Supports	23.40	22.68	0.00
Two Girders	25.81	24.39	0.00

Table 7. Effects of end skew angles

skew angle degree	M_x kN-m/m	M_y kN-m/m	M_{xy} kN-m/m
0	25.81	24.39	0.00
10	25.77	24.36	−0.18
20	25.63	24.25	−0.36
30	25.40	24.03	−0.51
40	25.09	23.72	−0.60

All the parameters used for this example, including the dimensions, material properties, loading and the mesh are the same as for the previous example, except the ends of the slab which are skewed at an angle θ as shown in Figure 23.

The resulting bending moments along line D-E-F are drawn in Figure 24. The maximum bending moments occur at the center of the loading area. The values of the bending moments for different skew angles are listed in Table 7. It can be seen that the skew angle of the slab ends has little influence on the local wheel effect.

5.7. COMBINED FE/BE ANALYSIS OF BOX GIRDER BRIDGES

If the girder have some irregularities such as openings or variable depth, then the girder can be modelled by finite element method, while the slab is still analyzed using the boundary element method described in the previous section. Thus, a combined finite element/boundary element analysis is performed.[12]

In this approach, the isoparametric shell elements[10] are employed. Each element has 4, 8 or 9 nodes (Figure 25). Along any straight edge of the element, the displacements are interpolated from the nodal values according to the Lagrange shape function (Equation 72), which is also used for the boundary elements. Therefore, a compatible connection between the boundary elements and finite elements along any interface can be achieved

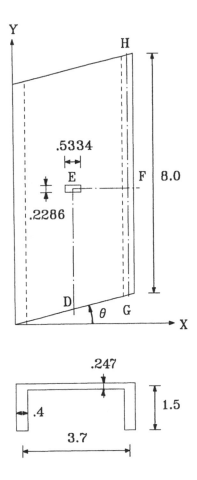

Figure 23. Skew slab on 2 girders (m).

if along the interface the boundary elements match the neighboring finite elements one to one. In other words, each boundary element on the interface should have three nodes if 8 or 9 node finite elements are used for girder, or two nodes if 4 node finite elements are chosen.

Once the mesh of finite elements is generated for girder, the overall stiffness matrix and the load vector are formed by following the commonly used procedures in finite element analysis. Then, by implementing the static condensation[12], the entire girder can be transformed into a substructure which only has four degrees of freedom at each interface node, namely

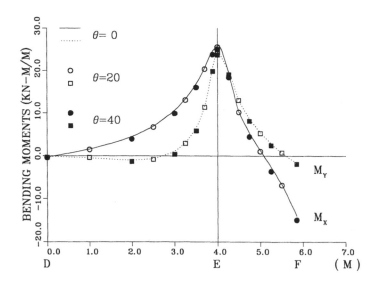

Figure 24. Bending moments along D-E-F.

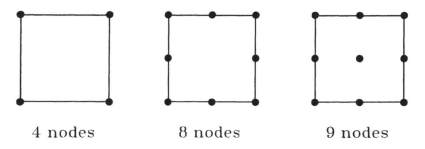

Figure 25. Isoparametric shell elements.

the displacements u, v, w and the normal rotation θ_n, with all other degrees of freedom being eliminated using Gauss elimination technique. Hence, the following matrix equation is obtained

$$[K]\{U_I^2\} = \{F_I^2\} + \{B^2\} \qquad (87)$$

where
    $[K]$ is the condensed stiffness matrix;
    U denotes the displacements and normal rotation at the interface nodes;

F is the vector of the unknown nodal forces;

B represents the effect of the external load on the substructure;

2 is the superscript identifying the girder substructure;

I is the subscript defining the interface between the slab and girder.

By taking into consideration the continuity conditions and the equilibrium conditions on each interface, as expressed in Equations 84 and 85 respectively, Equation 87 can be rewritten as:

$$[K]\{U_I^1\} = -[M](\{P_I^{1l}\} + \{P_I^{1r}\}) + \{B^2\} \tag{88}$$

where

1 is the superscript identifying the boundary element analysis;

P represents the nodal values of boundary forces and normal moment;

superscripts $1l$ and $1r$ refer respectively to the left and the right panels of the interface;

$[M]$ is the transformation matrix from the distributed boundary forces on the interface boundary elements to the nodal forces at the nodes of the finite elements. If node i and node j are in the same boundary element e, the transformation coefficient is

$$M_{ij} = \sum_e \int_e L_i(y) L_j(y)\, dy$$

The summation is necessary only if $i = j$ and i is the joint between two adjacent boundary elements. If nodes i and j are not located in the same element, then $M_{ij} = 0$.

Considering all the panels and girders, and using the Equation 81 and Equation 88 together, will yield enough equations for solving all the unknowns of the boundary elements, including those at the nodes on the interfaces. By backsubstituting into the girder substructures, all the unknown degrees of freedom can be obtained. Then, the displacements and stresses at any interested points can be readily calculated.

As a numerical example, a two-cell box girder bridge under truck loading is analyzed using the present approach.[12] This 20.0 m long, 8.4 m wide and 1.45 m deep two-cell box girder bridge is simply supported at both ends and loaded by four HS15-44 trucks[11], as shown in Figure 26. The load of each front wheel is uniformly distributed on a patch of 0.12 m width and 0.53 m length with intensity $P_1 = 285$ kN/m^2, while the load of each rear wheel is acting on a patch of 0.23 m width and 0.53 m length with intensity $P_2 = 590$ kN/m^2. The thickness of the top slab is 0.25 m, web 0.35 m and bottom flange 0.30 m. The material properties are $E = 3,200$ MPa and $\nu = 0.15$.

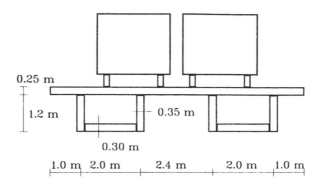

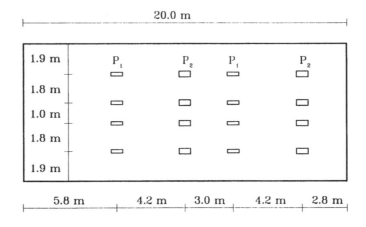

Figure 26. Two-cell box girder bridge under truck loading.

The boundaries of each slab panel are divided into a number of quadratic boundary elements with three nodes each. Eight elements are employed on each longitudinal side, while on each transverse side two elements are used for interior panels and one for overhangs. A 8 × 1 mesh using 8-node finite elements is used to model the webs and the bottom flanges. Details of the boundary element and finite element idealizations are given in Figure 27.

The deflection along the longitudinal center line of the top slab and the longitudinal bending moment along the longitudinal center line of the first wheel loads obtained from BEM-FEM and FSM solutions are presented in Tables 8 and 9. The maximum difference in the deflection and the moment between the two solutions are two and four percent respectively.

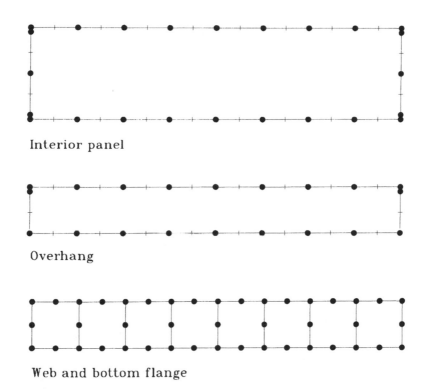

Interior panel

Overhang

Web and bottom flange

Figure 27. FE and BE idealizations.

Table 8. Deflection along longitudinal center line of slab

| Longitudinal | Deflection (mm) | |
coordinate (m)	FSM	BEM-FEM
2.0	6.93	6.84
4.0	12.73	12.48
5.8	16.97	16.64
8.0	20.44	20.23
10.0	22.36	21.97
13.0	19.13	18.92
16.0	12.74	12.54
17.2	9.62	9.43
18.0	6.93	6.83

Table 9. Longitudinal moment along the longitudinal center line of the wheel loads

Longitudinal coordinate (m)	Moment (kN.m/m)	
	FSM	BEM-FEM
2.0	1.749	1.783
4.0	0.875	0.849
5.8	1.452	1.396
8.0	0.626	0.613
10.0	10.360	0.962
13.0	1.442	1.389
16.0	0.874	0.857
17.2	5.172	4.974
18.0	1.749	1.778

References

1. Cheung, Y.K., 1976, Finite Strip Method in Structural Analysis (Pergaman Press, Oxford).
2. Cheung, M.S., W. Li and S.E. Chidiac, 1995, Finite Strip Analysis of Bridges (Chapman and Hall Ltd., London).
3. Cheung, M.S. and W. Li, 1991, *Computers and Structures*, **41**, 1119–1124.
4. Cheung, M.S. and W. Li, 1992, *Computers and Structures*, **45**, 1–7.
5. Cheung, M.S., G. Akhras and W. Li, 1994, *Journal of Structural Engineering*, ASCE, **120**, 716–727.
6. Timoshenko, S. and S. Woinowsky-Krieger, 1959, *Theory of Plates and Shells* (McGraw-Hill, New York), 2nd ed.
7. Brebbia, C.A. and J.J. Connor, 1973, *Fundamentals of Finite element Techniques* (Butterworths, London).
8. Abdel-Akher, A. and G.A. Hartley, 1989, *International Journal for Numerical Methods in Engineering*, **28**, 75–93.
9. Brebbia, C.A., J.C.F. Telles and L.C. Wrobel, 1984, *Boundary Element Techniques, Theory and Applications in Engineering* (Springer-Verlag, New York).
10. Cook, R.D., D.S. Malkus and M.E. Plesha, 1989, *Concepts and Applications of Finite Element Analysis* (John Wiley and Sons, New York): 3rd ed.
11. Rowe, R.E., 1976, *Concrete Bridge Design* (Applied Science Publishers Ltd., London).
12. Ezeedin M. Galuta, 1993, *Combined Boundary Element and Finite Element Analysis of Composite Bridges*, PhD thesis, Department of Civil Engineering, University of Ottawa, Ottawa, Canada.

6 NONLINEAR ANALYSIS OF THIN-WALLED STRUCTURES

F.G.A. ALBERMANI and S. KITIPORNCHAI

Department of Civil Engineering, University of Queensland, Brisbane Queensland, 4072 Australia

A finite element method capable of predicting the buckling capacity or the full nonlinear response of thin-walled structures under any general load and boundary conditions is presented. A rectangular thin plate element with 30 degrees of freedom is used. The linear and geometric stiffness matrices for this element are derived explicitly using symbolic manipulation, thereby eliminating the need for the expensive process of numerical integration. Further, the explicit form of the stiffness matrices makes it easier to interpret the physical significance of the various stiffness terms. Formex formulation is used for the automatic generation of the data necessary for the analysis. Numerical examples of thin-walled components are presented to demonstrate the accuracy and versatility of the method.

6.1. INTRODUCTION

Thin-walled structural members are used widely because of their relatively light weight, ease of fabrication, construction and availability in a variety of sectional shapes. These members are, however, highly susceptible to instability. Accurate determination of the buckling load incorporating all possible buckling modes including local, distortional and flexural-torsional buckling (see Figure 1) is therefore very important. One-dimensional beam-column modelling of thin-walled members is only capable of predicting accurately the flexural-torsional buckling modes.[1,2] Such a model cannot easily be used to predict the buckling load when local and distortional buckling modes are dominant. To predict these buckling modes, methods such as the finite strip[3,4] or finite element[5-7] methods are normally employed.

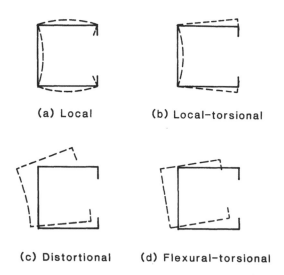

(a) Local   (b) Local-torsional

(c) Distortional   (d) Flexural-torsional

Figure 1. Possible buckling modes: (a) Local; (b) local-torsional; (c) distortional; (d) flexural–torsional.

In this Chapter, a rectangular thin plate element having 4-corner nodes with 7 degrees of freedom (d.o.f.) per node (i.e. 3 translations, 3 rotations and 1 warping) and 2 mid-edge nodes with a single translational d.o.f. per node is derived. The linear and geometric stiffness matrices are obtained explicitly using symbolic manipulation[8] which makes the element more efficient because numerical integration can be avoided. The explicit form of the stiffness matrices also makes it easier to interpret the physical significance of the various stiffness terms. This element is then implemented in a solution method for predicting either the various buckling modes (bifurcation analysis) or the full nonlinear response of thin-walled structures under general loading and boundary conditions. For members composed of rigid flanges and flexible webs where a distortional buckling mode is more dominant, a lower order plate element coupled with a beam-column element is used to form a super element to model the structural member.[9]

6.2. PROBLEM FORMULATION

Figure 2 shows a 6-node 30 d.o.f. rectangular thin plate element. The right hand orthogonal coordinate system x, y, z is chosen so that the x and y axes

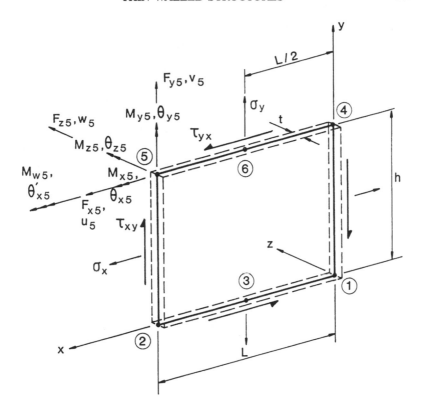

Figure 2. Thin plate element with generalized forces and stresses, displacements and local coordinate axes.

pass through the plate middle surface along the edge of the element. Seven actions (F_x, F_y, F_z, M_x, M_y, M_z and M_ω) with corresponding displacement components (u, v, w, θ_x, θ_y, θ_z and $\partial\theta_x/\partial x$) can be applied at each corner node of the thin plate element. In addition, a single action, F_x, with a corresponding displacement, u, can be applied at the mid-edge nodes along the x direction.

6.2.1. Strain Energy Functional

The strain energy functional, neglecting higher order terms, is given by

$$U = \frac{1}{2}\int_v (\underset{\sim}{\varepsilon}_L^T \, \mathbf{D} \, \underset{\sim}{\varepsilon}_L + 2\underset{\sim}{\varepsilon}_N^T \underset{\sim}{\varepsilon}_o)dv \tag{1}$$

in which ε_L and ε_N are the incremental linear and nonlinear components of Green-Lagrange strain tensor, $\sigma_o = \mathbf{D}\varepsilon_L$ is the Cartesian component of the Cauchy stress tensor, $\mathbf{D}$ is the material matrix and v is the volume of the continuum.[6,7] Using Kirchhoff's hypothesis of negligible transverse shear deformation, a line normal to the plate's undeformed middle surface is assumed to remain normal to the deformed middle surface. Therefore, the displacements $\tilde{u}$, $\tilde{v}$ and $\tilde{w}$ of an arbitrary point on the cross-section of the thin plate element (Figure 2) in the x, y and z directions due to in-plane stretching and out-of-plane bending can be written in terms of the middle surface in-plane and out-of-plane displacements, u_s, v_s and w_s,

$$\tilde{u} = u_s - z\frac{\partial w_s}{\partial x} \tag{2}$$

$$\tilde{v} = v_s - z\frac{\partial w_s}{\partial y} \tag{3}$$

$$\tilde{w} = w_s \tag{4}$$

Using the strain-displacement relation together with Equations (2)–(4), and neglecting higher order terms, the expression for the thin plate strain energy can be obtained

$$
\begin{aligned}
u = \int_v \Bigg[& D_{11}z^2\left(\frac{\partial^2 w_s}{\partial x^2}\right)^2 + (D_{12}+D_{21})z^2\left(\frac{\partial^2 w_s}{\partial x^2}\right)\left(\frac{\partial^2 w_s}{\partial y^2}\right) \\
& + D_{22}z^2\left(\frac{\partial^2 w_s}{\partial y^2}\right)^2 + 4D_{33}z^2\left(\frac{\partial^2 w_s}{\partial x \partial y}\right)^2 + D_{11}\left(\frac{\partial u_s}{\partial x}\right) \\
& + (D_{12}+D_{21})\left(\frac{\partial u_s}{\partial x}\right)\left(\frac{\partial v_s}{\partial y}\right) + D_{22}\left(\frac{\partial v_s}{\partial y}\right)^2 \\
& + D_{33}\left\{\left(\frac{\partial^2 u_s}{\partial y}\right)^2 + \left(\frac{\partial v_s}{\partial x}\right)^2 + 2\left(\frac{\partial u_s}{\partial y}\right)\left(\frac{\partial v_s}{\partial x}\right)\right\} \\
& + \sigma_{xx}\left(\frac{\partial^2 w_s}{\partial x}\right) + \sigma_{yy}\left(\frac{\partial w_s}{\partial y}\right) + 2\tau_{xy}\left(\frac{\partial w_s}{\partial x}\right)\left(\frac{\partial w_s}{\partial y}\right) \\
& + \sigma_{xx}\left\{\left(\frac{\partial u_s}{\partial x}\right)^2 + \left(\frac{\partial v_s}{\partial x}\right)^2\right\} + \sigma_{yy}\left\{\left(\frac{\partial u_s}{\partial y}\right)^2\left(\frac{\partial v_s}{\partial y}\right)\right\} \\
& + 2\tau_{xy}\left\{\left(\frac{\partial u_s}{\partial x}\right)\left(\frac{\partial u_s}{\partial y}\right) + \left(\frac{\partial v_s}{\partial x}\right)\left(\frac{\partial v_s}{\partial y}\right)\right\}\Bigg] dv
\end{aligned}
\tag{5}
$$

in which E is the Young's modulus, v the Poisson's ratio and

$$D_{11} = D_{22} = \frac{E}{1 - v^2} \tag{6a}$$

$$D_{12} = D_{21} = \frac{vE}{1 - v^2} \tag{6b}$$

$$D_{33} = \frac{e}{2(1 + v)} \tag{6c}$$

6.2.2. Stress Fields

Since explicit stiffness matrices are desired, the stress components have to be approximated in terms of the element stress resultants. The direct stress, σ_{xx}, can be approximated by a linear function along the x and y directions (see Figure 2) to give

$$
\begin{aligned}
\sigma_{xx} = & \frac{F_{x5} + F_{x2}}{ht} \rho_2 - \frac{F_{x4} + F_{x1}}{ht} \rho_1 \\
& + \left[\frac{(F_{x1} - F_{x4})h}{2} \rho_1 \frac{(F_{x2} - F_{x5})h}{2} \rho_2 \right] \frac{\bar{y}}{I_2} + \sigma_r
\end{aligned}
\tag{7}
$$

in which

$$\bar{y} = y - \frac{h}{2} \tag{8a}$$

$$I_z = \frac{h^3 t}{12} \tag{8b}$$

$$\rho_1 = 1 - \frac{x}{L} \tag{8c}$$

$$\rho_2 = \frac{x}{L} \tag{8d}$$

and σ_r is the residual stress in the cross-section resulting from fabrication, h, L and t are the width, length and thickness of the thin plate element. The stresses, σ_{yy} and τ_{xy}, may be approximated by

$$\sigma_{yy} = \frac{F_{y4} + F_{y5}}{LT} \tag{9}$$

$$\tau_{xy} = \frac{F_{y5} + F_{y2}}{ht} \tag{10}$$

6.2.3. Displacement Fields

The displacement field, u_s, in Equation (5) can be approximated using linear and quadratic interpolation polynomial functions, N_1 and f_2, in the y and x directions, respectively, i.e.

$$U_s = \mathbf{N}_1 \begin{bmatrix} \mathbf{f}_2 & 0 \\ 0 & \mathbf{f}_2 \end{bmatrix} \qquad (11)$$

in which

$$U_e = \langle u_1 u_3 u_2 u_4 u_6 u_5 \rangle^T \qquad (12a)$$

$$\mathbf{N}_1 = \langle \xi_1 \xi_2 \rangle \qquad (12b)$$

$$\mathbf{f}_2 = \langle \rho_1 (1 - 2\rho_2) 4\rho_1 \rho_2 - \rho_2 (\rho_1 - \rho_2) \rangle \qquad (12c)$$

$$\xi_1 = 1 - \frac{y}{h}; \, \xi_2 = \frac{y}{h} \qquad (13)$$

Values of ρ_1 and ρ_2 are given in Equations (8c) and (8d). The displacement field, v_s, in Equation (5) can be approximated using linear and Hermitian interpolation polynomial functions, $\mathbf{N}_1$ and $\mathbf{f}_3$, in the y and x directions, respectively. It can be expressed as

$$v_s = \mathbf{N}_1 \begin{bmatrix} \mathbf{f}_3 & \mathbf{0} \\ \mathbf{0} & \mathbf{f}_3 \end{bmatrix} \mathbf{v}_e \qquad (14)$$

in which

$$\mathbf{v}_e = \langle v_1 \theta_{z1} v_2 \theta_{z2} v_4 \theta_{z4} v_5 \theta_{z5} \rangle^T \qquad (15a)$$

$$\mathbf{f}_3 = \langle (3 - 2\rho_1)\rho_1^2 \rho_1^2 \rho_2^L (3 - 2\rho_2)\rho_2^2 - \rho_1 \rho_2^2 L \rangle \qquad (15b)$$

According to Equation (14), a linear shape function is used for the displacement field, v_s, along the y direction. For general in-plane loading, this may increase the in-plane bending strain energy resulting in an over-stiff behavior. Wood and Zienkewicz[10] suggested a remedy to this problem by modifying the material matrix ($D_{11} = D_{22} = E$ and $D_{12} = D_{21} = 0$) for the in-plane stiffness. This procedure has been adopted in this work. Further, the difference in interpolating different stress components (Equations (7), (9) and (10)) makes the present element more suitable for modelling structural members where the longitudinal stress σ_{xx} is the dominant stress.

Finally, Hermitian interpolation polynomial functions $\mathbf{f}_3$ and $\mathbf{N}_3$ are used to approximate the displacement field, w_s, in the x and y directions, respectively, i.e.

$$\omega_s = \mathbf{N}_3 \begin{bmatrix} \mathbf{f}_3 & 0 & 0 & 0 \\ 0 & \mathbf{f}_3 & 0 & 0 \\ 0 & 0 & \mathbf{f}_3 & 0 \\ 0 & 0 & 0 & \mathbf{f}_3 \end{bmatrix} \mathbf{w}_e \qquad (16)$$

in which

$$\mathbf{w}_e = \langle w_1 - \theta_{y1}w_2 - \theta_{y2}\theta_{x1}\frac{\partial \theta_{x1}}{\partial x}\theta_{x2}\frac{\partial \theta_{x2}}{\partial x}$$
$$w_4 - \theta_{y4}w_5 - \theta_{y5}\theta_{x4}\frac{\partial \theta_{x4}}{\partial x}\theta_{x5}\frac{\partial \theta_{x5}}{\partial x}\rangle^T \tag{17a}$$

$$\mathbf{N}_3 = \langle (3 - 2\xi_1)\xi_1^2\xi_1^2\xi_2 h(3 - 2\xi_2)\xi_2^2 - \xi_1\xi_2^2 h\rangle \tag{17b}$$

Values of ξ_1, ξ_2 and $\mathbf{f}_3$ are given in Equations (13) and (15b), respectively. The inclusion of $\partial \theta_x / \partial x$ in Equation (17a) is to retain compatibility with the thin-walled beam-column element when these two elements are combined to form a super element.[9]

6.2.4. Stiffness Matrices

Substituting Equations (7), (9)–(11), (14) and (16) into Equation (5) and integrating using symbolic manipulation[8], the equilibrium of the thin plate element can be expressed as

$$(\mathbf{k}_L + \mathbf{k}_G)\mathbf{r} = \mathbf{f} \tag{18}$$

in which

$$\mathbf{r} = \langle u_1 v_1 w_1 \theta_{x1}\theta_{y1}\theta_{z1}\frac{\partial \theta_{x1}}{\partial x}u_2 v_2 w_2 \theta_{x2}\theta_{y2}\theta_{z2}\frac{\partial \theta_{x2}}{\partial x}u_3 u_4 v_4 w_4 \theta_{x4}\theta_{y4}\theta_{z4}\frac{\partial \theta_{x4}}{\partial x}$$
$$u_5 v_5 w_5 \theta_{x5}\theta_{y5}\theta_{z5}\frac{\partial \theta_{x5}}{\partial x}u_6 \rangle^T \tag{19a}$$

$$\mathbf{f} = \langle F_{x1}F_{y1}F_{z1}M_{x1}M_{y1}M_{z1}M_{\omega 1}F_{x2}F_{y2}F_{z2}M_{x2}M_{y2}M_{z2}M_{\omega 2}F_{x3}F_{x4}$$
$$F_{y4}F_{z4}M_{x4}M_{y4}M_{z4}M_{\omega 4}F_{x5}F_{y5}F_{z5}M_{x5}M_{y5}M_{z5}M_{\omega 5}F_{x6} \rangle^T \tag{19b}$$

The linear, $\mathbf{k}_L$, and geometric, $\mathbf{k}_G$, stiffness matrices for the thin plate element may thus be obtained explicitly. These matrices are given in Appendix 2 at the end of this Chapter.

6.2.5. Bifurcation Analysis

The buckling load can be obtained from the non-trivial solution of

$$\det(\mathbf{K}_L + \lambda_{cr}\mathbf{K}_G) = 0 \tag{20}$$

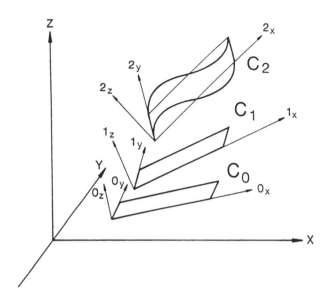

Figure 3. Element Deformations.

in which λ_{cr} is the buckling load factor and $\mathbf{K}_L$ and $\mathbf{K}_G$ are the linear and geometric stiffness matrices for the structure which can be obtained from transformation and assembly of the element matrices.[6] To extract the buckling load factor, λ_{cr}, from Equation (20), a bisection algorithm is used.

6.3. FULL NONLINEAR ANALYSIS

Figure 3 shows the deformation path of an element which may be described using three configurations: C_0, C_1 and C_2. Configuration C_0 represents the initial undeformed state; C_1 the current (known) deformed state and C_2 a neighboring or desired deformed state. The following nomenclature has been adopted[11]; a left superscript denotes the configuration in which a quantity occurs, while a left subscript denotes to the configuration in which a quantity is measured.

Adopting an updated Lagrangian frame of reference and using the small strain hypothesis[11], the incremental equilibrium equation at the element level is described as

$$({}_1\mathbf{k}_L + {}_1\mathbf{k}_G)^2_1 \Delta\mathbf{r} = {}^2_1\Delta\mathbf{f} \tag{21}$$

in which $^2_1\Delta\mathbf{r}$ and $^2_1\Delta\mathbf{f}$ are the incremental nodal displacement and force vectors between the C_1 and C_2 states, respectively, and $_1\mathbf{k}_L$ and $_1\mathbf{k}_G$ are the linear and geometric stiffness matrices measured in configuration C_1. The left subscripts for the stiffness matrices will be dropped for clarity. Note that all of these quantities are defined in the local coordinate system.

In order to determine the full nonlinear response, an incremental-iterative solution method is used. This method involves three principal stages. These are: (i) Predictor stage: evaluating overall structural stiffness and solving for the incremental displacements from the approximated incremental equilibrium equation for the structure; (ii) Corrector stage: determining the exact nodal forces for each element using particular force recovery algorithms; and (iii) Detector stage: checking the equilibrium condition between internal and applied forces for the structure in the C_2 deformed state to see whether further iteration is required.

6.3.1. Predictor Stage

At the beginning of each load cycle, the linear and geometric stiffness matrices, $\mathbf{k}_L$ and $\mathbf{k}_G$, are formed for each element. The tangent stiffness matrix, $\mathbf{k}_{Te}$, is obtained for each element by augmenting $\mathbf{k}_L$ by $\mathbf{k}_G$. This matrix is then transformed from the local coordinate system to the global coordinate system to obtain $\mathbf{k}_{Tg}$, and assembled to form the overall structural stiffness matrix, $\mathbf{K}_T$.[6,7] The overall structural stiffness matrix, $\mathbf{K}_T$, is factorized and solved for the incremental nodal displacements, $^2_1\Delta\mathbf{D}$, in the global coordinate system. Since the full Newton-Raphson method has been adopted in this study, the assembly and factorization procedures have been repeated at every iteration during the load cycle. Mathematically, the described procedure can be summarized as follows:

$$\mathbf{k}_{Te} = \mathbf{k}_L + \mathbf{k}_G \tag{22}$$

$$\mathbf{k}_{Tg} = {}^1\mathbf{L}^T\mathbf{k}_{Te}\,{}^1\mathbf{L} \tag{23}$$

$$\mathbf{K}_T = \sum_{i=1}^{n}\mathbf{k}_{Tg} \tag{24}$$

$$\mathbf{K}_T\,{}^2_1\Delta\mathbf{d} = {}^2_1\Delta\mathbf{f}_{\text{ext}} \tag{25}$$

in which n is the total number of elements in the structure, $^2_1\Delta\mathbf{f}_{\text{ext}}$ is the vector of the applied force increments and $^1\mathbf{L}$ is the transformation matrix in C_1 for each element (see Appendix 1). In the updated Lagrangian approach, the transformation matrix is continuously updated for each new configuration of the element.

6.3.2. Corrector Stage

The global incremental nodal displacements, ${}_1^2\Delta\mathbf{r}_g$, for each element extracted from vector ${}_1^2\Delta\mathbf{d}$ are transformed back to the local coordinate system to obtain vector ${}_1^2\Delta\mathbf{r}$ as follows:

$${}_1^2\Delta\mathbf{r} = {}^1\mathbf{L}_1^2\Delta\mathbf{r}_g \tag{26}$$

In the corrector stage, the force recovery calculation for each element has to be exact, at least in the limit sense, whereas the process in the predictor stage can be approximated. In this study, it has been found that even if $\mathbf{k}_G$ is not included in Equation (22), the solution will converge to the correct answer but with more iterations. This highlights the fact that the most important part of the solution process is the proper expression of the equilibrium condition at the deformed state. In this work, the following force recovery algorithm has been adopted

$${}^2\mathbf{f} = {}^1\mathbf{f} + (\mathbf{k}_L + \mathbf{k}_G)_1^2\Delta\mathbf{r}_{def} \tag{27}$$

in which ${}^2\mathbf{f}$ is the element nodal resisting force vector in C_2 referred to its local coordinate system, and ${}_1^2\Delta\mathbf{r}_{def}$ is the incremental nodal deformation for an element which can be obtained from ${}_1^2\Delta\mathbf{r}$ after eliminating the rigid body motions (see Figure 4). The rigid body motions can be expressed in terms of the element nodal displacements as follows:

$$\theta_{xr} = \sin^{-1}\left(\frac{-w_1 - w_2 + w_4 + w_5}{2h}\right) \tag{28}$$

$$\theta_{yr} = \sin^{-1}\left(\frac{w_1 - w_2 + w_4 - w_5}{2L}\right) \tag{29}$$

$$\theta_{zr} = \frac{1}{2}\left[\sin^{-1}\left(\frac{u_1 + u_2 - u_4 - u_5}{2h}\right) + \sin^{-1}\left(\frac{-v_1 + v_2 - v_4 + v_5}{2L}\right)\right] \tag{30}$$

$$u_r = \frac{1}{4}(u_1 + u_2 + u_4 + u_5) - \frac{\eta h}{2}\sin(\theta_{zr}) \tag{31}$$

$$v_r = \frac{1}{4}(v_1 + v_2 + v_4 + v_5) + \frac{\xi L}{2}\sin(\theta_{zr}) \tag{32}$$

$$w_r = \frac{1}{4}(w_1 + w_2 + w_4 + w_5) + \frac{\eta h}{2}\sin(\theta_{xr}) - \frac{\xi L}{2}\sin(\theta_{yr}) \tag{33}$$

in which

$$\eta = \frac{2y}{h} - 1; \quad \text{and} \quad \xi = \frac{2x}{L} - 1 \tag{34}$$

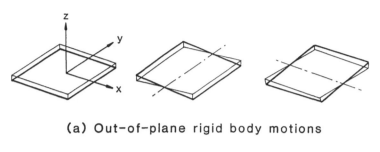

(a) Out-of-plane rigid body motions

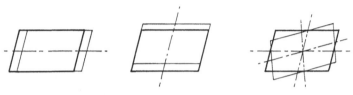

(b) In-plane rigid body motions

Figure 4. Rigid Body Motion.

The reason for using the incremental nodal deformations, $^2_1\Delta r_{\text{def}}$, instead of the incremental nodal displacements, $^2_1\Delta r$, is to eliminate the fictitious forces which may be generated by the stiffness matrices due to large rigid body motions. The resulting element nodal resisting force vector in configuration C_2, ^2f, is obtained without the need for transformation from C_1 to C_2.

6.3.3. Detector Stage

The element nodal resisting force vector, ^2f, must be transformed to the global coordinate system to obtain the element nodal resisting force vector, ^2f_e, in the global coordinate system:

$$^2f_e = {}^2L^{T}{}^2f \tag{35}$$

The element nodal resisting force vector, ^2f_e, in the global coordinate system is then assembled to obtain the total resisting force vector, $^2f_{\text{int}}$, of the structure:

$$^2f_{\text{int}} = \sum_{i=1}^{n} {}^2f_e \tag{36}$$

The out-of-balance force vector, $\mathbf{q}_r$, can be found from the difference between the total applied force vector, $^2\mathbf{f}_{ext}$, and the total resisting force vector, $^2\mathbf{f}_{int}$:

$$\mathbf{q}_r = {}^2\mathbf{f}_{ext} - {}^2\mathbf{f}_{int} \tag{37}$$

In order to satisfy the equilibrium condition, the out-of-balance forces must be dissipated through an iterative procedure. If $\|\mathbf{q}_r\|$ is smaller than a certain norm measure then the iteration process stops and the solution proceeds to the next load cycle.

6.4. DATA PROCESSING USING FORMEX ALGEBRA

One of the first steps in the analysis of a structure is the generation of the data containing the necessary information regarding the elements of the structure, their connectivity, geometry, loading and support conditions. Without a suitable conceptual tool, this task can be very tedious, time consuming and susceptible to error. Formex algebra[12,13] provides a convenient technique for data generation. The formex algebra is a mathematical system that consists of a set of abstract objects, known as formices, and a number of rules according to which these objects may be manipulated. In this application, a single plate element can be viewed as a formex and used as a generic element for the rest of the structure (see Figure 5(a)). Once this formex is represented explicitly, certain functions are used to generate the structure. These functions perform the necessary propagation, curtailment and projection tasks. Once the structure topology is generated using a convenient reference system, the actual geometry of the configuration is then imposed using a suitable geometric transformation.

This technique is implemented as a preprocessor to the analysis program. The program is implemented on a personal computer. Prior to the analysis, the data is generated automatically and viewed for checking. Figure 5(b) and (c) show some finite element meshes generated by this technique.

6.5. NUMERICAL EXAMPLES

In order to validate this finite element formulation, buckling solutions and the full nonlinear response for a number of example problems have been obtained. Where possible, the accuracy of the finite element method is demonstrated by comparing the derived solutions with test results and with other independent published solutions based on different numerical techniques.

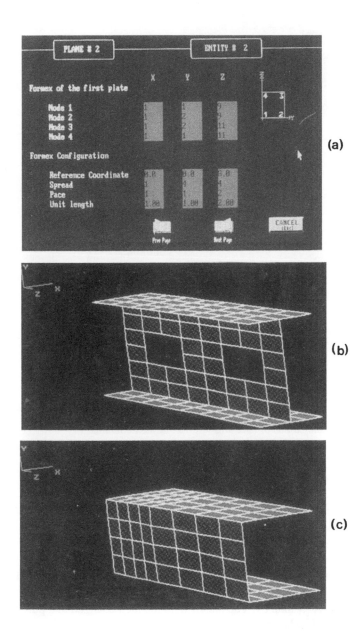

Figure 5. Formex configuration processing. (See Color Plate I)

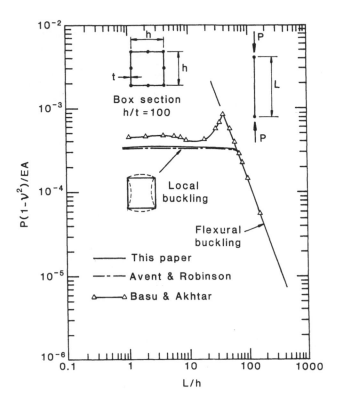

Figure 6. Comparison of buckling solution for box columns.

6.5.1. Buckling of Box Section Columns

In this example, the stability characteristic of a simply supported square box column with $v = 0.3$ and $h/t = 100$ (see Figure 6) has been investigated. This problem was studied by Avent and Robinson[14] who solved the governing differential equations by representing the various displacement functions with infinite Fourier series. Avent and Robinson found that the overall flexural buckling mode for the column with $h/t = 100$ occurred when the L/h ratio was greater than 67. The buckling mode then changed to a purely local buckling mode when the L/h ratio was smaller than 67.

Recently, this problem was re-examined by Basu and Akhtar[15] who used the p-version finite element method to obtain the buckling load. Basu and Akhtar's results are compared with Avent and Robinson's in Figure 6. It is seen that there is some disagreement in the two sets of results. Basu and Akhtar predicted a lower L/h ratio when local buckling occurs and a higher overall local buckling load.

The problem has been analyzed using the proposed finite element formulation. Because of symmetry, only half the column length needed to be analyzed. In the analysis, each side of the box width was divided into two segments (see Figure 6). Each segment along the column half length was divided into 3 to 20 divisions, depending on the column slenderness. More elements were used for longer columns with a short buckled wave length.

The present results are compared with Avent and Robinson[14] and Basu and Akhtar[15] in Figure 6. Solutions obtained by the authors are in excellent agreement with Avent and Robinson's. The higher buckling load predicted by Basu and Akhtar may have been due to improper modelling of the boundary conditions and to the use of a lower order polynomial to model the longitudinal displacements.

6.5.2. Buckling of Lipped Channel Beams

6.5.2.1. Seah and Khong's experiments

An experimental investigation was carried out by Seah and Khong[16] to study the elastic buckling behavior of simply supported channel beams (Figure 7(a)) having different outer lip widths bent about the minor principle axis (lips bent outwards). The test beams, 1920 mm span, were loaded by two equal concentrated loads at the bottom edge level (see Figure 7(a)), acting at equal distances of 160 mm from the end supports. Tested specimens were manufactured by cold-forming galvanized mild steel sheet ($E = 2 \times 10^5$ MPa, $\nu = 0.25$ and average yield stress = 286.72 MPa) to normal engineering tolerances. The geometry and dimensions of the test beams ($h_2/h_1 \cong 0.8$) are shown in Figure 7 and summarized in Table 1. The experimental buckling moments of the test beams are shown in Table 1 and Figure 8.

The test beams have been analyzed using the finite element formulation. Because of symmetry, only half the beam length needed to be analyzed. The channel section was discretized as shown in Figure 7(b), with half the beam length divided into six segments. The total number of elements used in modelling the half beam length was thus 60 for the unlipped channel beams and 84 for the lipped channel beams. The boundary and loading conditions at the bottom edge of the section were modelled as shown in Figure 7(b).

Theoretical predictions from the authors' finite element method are compared with Seah and Khong's experimental results in Table 1 and Figure 8. The predicted results indicate that Test Beams T1 to T5 are controlled by

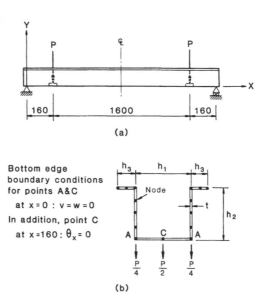

(b)

Figure 7. Lipped channel beams.

flexural-torsional buckling while Test Beam T6 failed in the local-torsional mode. The authors' predicted buckling moments are generally in good agreement with experimental results. The discrepancies in Test Beam T6, which was controlled by local-torsional buckling, may have been associated with the fact that the post-local buckling strength is not incorporated in the present bifurcation analysis.

Table 1. Comparison of predicted and experimental buckling moments of lipped channel beams

Specimen	Dimension				Buckling Moment (kN-mm)		Percent Difference
	t (mm)	h_1 (mm)	h_2 (mm)	h_3 (mm)	Test (a)	Finite Element (b)	
T1	0.75	50.25	39.25	14.125	163.68	159.20	−2.74
T2	0.76	51.24	40.24	14.620	161.98	169.51	4.65
T3	0.76	49.24	41.24	14.620	145.98	167.87	14.99
T4	0.75	51.25	40.25	10.625	142.40	161.78	13.61
T5	0.76	51.24	40.74	6.620	172.66	168.93	−2.16
T6	0.75	50.25	39.25	0.000	72.98	59.46	−18.53

(a) Seah and Khong[16]; (b) Authors

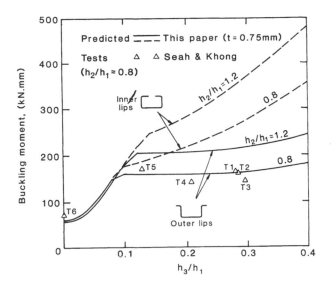

Figure 8. Comparison of buckling moments for lipped channel beams.

6.5.2.2. Effect of lip stiffener on buckling moment

The finite element method has been used to study the effect of the orientation and length of lip stiffeners on the buckling characteristic of the test channel beam. Buckling moments have been calculated for channel beams with h_2/h_1 ratios of 0.8 and 1.2, with lips bent either inwards or outwards (see Figure 8). The buckling moments of channel beams with varying outer or inner lip length ratios, h_3/h_1, are compared in Figure 8. In general, an inner lipped channel beam is stronger than an equivalent outer lipped channel beam because inner lipped channel sections have greater resistance to warping deformations.

The buckling curves shown in Figure 8 exhibit distinct discontinuities which correspond to changes in the buckling modes as the lip length ratio, h_3/h_1, increases. For a section with a given h_2/h_1 ratio, the three buckling modes which may occur are illustrated in Figure 9. For a section with short lip length (i.e. h_3/h_1 small), the beam buckles in the local-torsional buckling mode (Figure 1(b)). As the lip length ratio increases, the buckling mode changes to distortional buckling mode (Figure 1(c)). For sections with sufficiently long lip length (i.e. h_3/h_1 large), the buckling mode is the flexural-torsional buckling mode (Figure 1(d)).

It is interesting to note that in the local-torsional buckling mode, increases in the h_2/h_1 ratio lead to a slightly lower buckling moment. Further, the

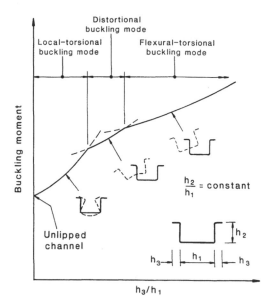

Figure 9. Buckling modes in lipped channel beams.

local-torsional buckling strengths of inner and outer lipped channel beams do not differ appreciably (see Figure 8). However, as the lip length ratio increases, the difference in the buckling strength becomes more pronounced. For large h_3/h_1 ratios (long lips) where the flexural-torsional buckling mode occurs, an inner lipped channel beam is considerably stronger than an equivalent outer lipped channel beam as seen in Figure 8. The flexural-torsional buckling strength for outer lipped channel beams is less sensitive to any increase in the lip length ratio h_3/h_1.

6.5.3. Square Plate Subjected to Edge Compressive Loads

A simply supported square plate loaded by a uniformly distributed in-plane loading on its two longitudinal edges is shown in Figure 10. The transverse in-plane displacements of the unloaded edges are restrained. The linear buckling load[17], P_{cr}, can be expressed in terms of

$$P_{cr} = \frac{k_c \pi^2 E t^3}{12h(1 - v^2)} \tag{38}$$

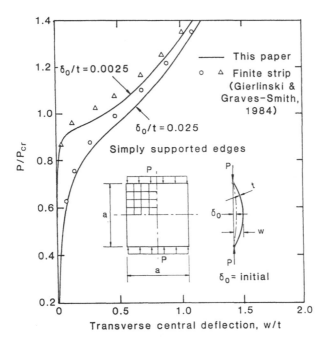

Figure 10. Load-deflection curves of a simply-supported plate under uniformly distributed in-plane load.

where $k_c = 4.88$. This is higher than in the case of a square plate with unloaded edges which are free to move transversely, in which $k_c = 4.00$. The post-buckling behavior can be initiated by introducing a small load imperfection into the structure. In this case couples, $P\delta_o$, were introduced at both of the longitudinal edges (see Figure 10), where e is the eccentricity of the applied compressive loads. Using a 4×4 mesh discretization for one-quarter of the structure, the nonlinear load-deflection curves for various eccentricities have been calculated as shown in Figure 10. A comparison has been made with the finite strip results of Gierlinski and Graves-Smith.[18] The finite strip results appear to be somewhat stiffer in the post-buckling range.

6.5.4. Hinged Cylinder With a Center Point Load

The hinged cylindrical shell shown in Figure 11 is a widely used test problem for limit point solution algorithms and large rotation shell finite element

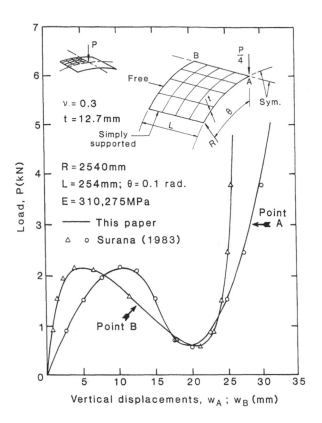

Figure 11. Snap-through of a shallow shell.

formulations. The longitudinal boundaries are hinged and immovable whereas the curved edges are completely free. This problem has been studied by Sabir and Lock[19] and more recently by Surana.[20] The proposed method was applied using a 4×4 mesh to discretize one-quarter of the structure. As shown in Figure 11, the results using the proposed finite element compare quite well with those of Surana.[20]

6.6. CONCLUSIONS

A versatile finite element method using a 6-node, 30 d.o.f. thin plate element capable of predicting the local, distortional and flexural-torsional buckling capacities of thin-walled structural members under any loading and boundary conditions has been presented. The linear and geometric stiffness matrices for this element have been obtained explicitly using symbolic manipulation, hence eliminating the need for the time-consuming process of numerical integration and facilitating the interpretation of various stiffness terms.

Based on the updated Lagrangian formulation and small strain hypothesis, this element can accurately predict the nonlinear pre- and post-buckling load-deflection paths of thin-walled plate structures involving cross-section distortion and local instability. The necessary data for the analysis is generated automatically using formex algebra. This technique is implemented as a preprocessor to the analysis program.

The proposed finite element formulation has been applied to a wide range of example problems using either a bifurcation analysis or full nonlinear analysis involving large displacements. It has been demonstrated that in all cases, the method is able to predict the structural response accurately.

References

1. Chan, S.L. and S. Kitipornchai, 1987, *Engng. Struct.*, **9**, 243–254.
2. Al-Bermani, F. and S. Kitipornchai, 1990, *J. Struct. Engn.*, ASCE, **116**(1), 215–234.
3. Plank, R.J. and W.H. Wittrick, 1974, *Int. J. Num. Methods Engng.*, **8**(2), 323–339.
4. Van Erp, G.M. and C.M. Menken, 1990, *Comm. Appl. Num. Methods*, **6**(6), 477–484.
5. Johnson, C.P. and K.M. Will, 1974, *J. Struct. Div.*, ASCE, **100**(ST3), 669–685.
6. Chin, C.K., Al-Bermani, F. and S. Kitipornchai, 1993, *J. Struct. Engng.*, ASCE, **119**(4), 1048–1068.
7. Chin, C.K., Al-Bermani, F. and S. Kitipornchai, 1994, *Int. J. Num. Methods Engng.*, **37**, 1697–1711.
8. Wolfram, S., 1991, *Mathematica* (Addison-Wesely, N.Y.), 2nd ed.
9. Chin, C.K., Al-Bermani, F. and S. Kitipornchai, 1992, *Engng Struct.*, **14**(2), 121–132.
10. R.D., R.D. and O.C. Zienkiewicz, 1977, *Comput. Struct.*, **7**(6), 725–735.
11. Bathe, K.J. and S. Bolourchi, 1980, *Comput. Struct.*, **11**(1/2), 23–48.
12. Nooshin, H., 1984, *Formex configuration processing in structural engineering* (Elsevier Applied Science Publishers).
13.. Chin, S. and F. Albermani, 1994, *Int. J. Space Structures*, **9**(3), 147–152.
14. Avent, R.R. and J.H. Robinson, 1976, *J. Struct. Div.*, ASCE, **102**(ST5), 1015–1029.
15. Basu, P.K. and M.N. Akhtar, 1991, *Thin-Walled Struct.*, **12**(4), 335–352.
16. Seah, L.K. and P.W. Khong, 1990, *J. Const. Steel Res.*, **17**(4), 265–282.
17. Timoshenko, S.P. and J.M. Gere, 1961, *Theory of elastic stability* (McGraw-Hill, New York), 2nd ed.
18. Gierlinski, J.T. and T.R. Graves-Smith, 1984, *Thin-Walled Struct.*, **2**(1), 27–50.
19. Sabir, A.B. and A.C. Lock, 1973, *Variational Methods in Engineering* (Southampton Univ. Press, eds: Brebia and Tottenham), Chap. 7.
20. Surana, K.S., 1983, *Int. J. Num. Meth. Eng.*, **19**(4), 581–615.

6.7. APPENDIX 1

6.7.1. Transformation Matrices

Each element stiffness matrix, k_{Te}, which is derived in local coordinate system (x, y and z) has to be transformed into the global coordinate system (X, Y and Z) before assembling with other elements of the structure into an overall structural stiffness matrix (Figure 3). The transformation method must be capable of dealing with any orientation or combination of elements in a three dimensional space.

Since the element can be described as a plane, the direction cosines of the normal to the plane (z-axis) can be written as:

$$\ell_{zX} = \frac{A_z}{\sqrt{(A_z^2 + B_z^2 + C_z^2)}}$$

$$\ell_{zY} = \frac{B_z}{(A_z^2 + B_z^2 + C_z^2)}$$

$$\ell_{zZ} = \frac{C_z}{\sqrt{(A_z^2 + B_z^2 + C_z^2)}}$$

The directional vector $\langle A_z, B_z, C_z \rangle$ is obtained from the cross product of the vector of node 1 to node 2 and the vector of node 1 to node 5 (Figure 2), in which

$$A_z = (Y_2 - Y_1)(Z_5 - Z_1) - (Y_5 - Y_1)(Z_2 - Z_1)$$
$$B_z = -(X_2 - X_1)(Z_5 - Z_1) + (X_5 - X_1)(Z_2 - Z_1)$$
$$C_z = (X_2 - X_1)(Y_5 - Y_1) - (X_5 - X_1)(Y_2 - Y_1)$$

The direction cosines of the y-axis are given by:

$$\ell_{yX} = \frac{A_y}{\sqrt{(A_y^2 + B_y^2 + C_y^2)}}$$

$$\ell_{yY} = \frac{B_y}{(A_y^2 + B_y^2 + C_y^2)}$$

$$\ell_{yZ} = \frac{C_y}{\sqrt{(A_y^2 + B_y^2 + C_y^2)}}$$

The directional vector $\langle A_y, B_y, C_y \rangle$ is obtained from the cross product of the directional vector $\langle A_z, B_z, C_z \rangle$ and the vector of node 1 to node 2, where

$$A_y = B_z(Z_2 - Z_1) - C_z(Y_2 - Y_1)$$
$$B_y = C_z(X_2 - X_1) - A_z(Z_2 - Z_1)$$
$$C_y = A_z(Y_2 - Y_1) - B_z(X_2 - X_1)$$

In a similar manner, the direction cosines of the x-axis are given by:

$$\ell_{xX} = \frac{A_x}{\sqrt{(A_x^2 + B_x^2 + C_x^2)}}$$

$$\ell_{xY} = \frac{B_x}{(A_x^2 + B_x^2 + C_x^2)}$$

$$\ell_{xZ} = \frac{C_x}{\sqrt{(A_x^2 + B_x^2 + C_x^2)}}$$

where

$$A_x = B_y C_z - C_y B_z$$
$$B_x = C_y A_z - A_y C_z$$
$$C_x = A_y B_z - B_y A_z$$

Therefore, the element stiffness matrix expressed in global coordinates is given by:

$$(\mathbf{k}_L + \mathbf{k}_G)_{\text{global}} = \mathbf{L}^T (\mathbf{k}_L + \mathbf{k}_G)_{\text{local}} \mathbf{L}$$

where

$$
\mathbf{L} =
\begin{bmatrix}
\underset{\sim}{\ell} & & & & & \\
 1 & & & zero & & \\
 & \underset{\sim}{\ell} & & & & \\
 & & 1 & & & \\
 & & & 1 & & \\
 & & & & \underset{\sim}{\ell} & \\
 & zero & & & 1 & \\
 & & & & & \underset{\sim}{\ell} \\
 & & & & & 1 \\
 & & & & & & 1
\end{bmatrix}
$$

and

$$
\underset{\sim}{\ell} =
\begin{bmatrix}
\ell_{xx} & \ell_{xY} & \ell_{xZ} & 0 & 0 & 0 \\
\ell_{yX} & \ell_{yY} & \ell_{yZ} & 0 & 0 & 0 \\
\ell_{zX} & \ell_{zY} & \ell_{zZ} & 0 & 0 & 0 \\
0 & 0 & 0 & \ell_{xX} & \ell_{xY} & \ell_{xZ} \\
0 & 0 & 0 & \ell_{yX} & \ell_{yY} & \ell_{yZ} \\
0 & 0 & 0 & \ell_{zX} & \ell_{zY} & \ell_{zZ}
\end{bmatrix}
$$

6.8. APPENDIX 2

The upper triangle of the linear stiffness matrix for the thin plate element k_L:

```
KL(1,1)=DD*7./9.*HW/L+G*2./15.*L/HW
KL(1,2)=G/20.
KL(1,6)=-G*7./120.*L
KL(1,8)=DD*HW/(9.*L)-G*L/(30.*HW)
KL(1,9)=-G/20.
KL(1,13)=G*L/40.
KL(1,15)=-DD*8./9.*HW/L+G*L/(15.*HW)
KL(1,16)=DD*7./18.*HW/L-G*2./15.*L/HW
KL(1,17)=G/20.
KL(1,21)=-G*7./120.*L
KL(1,23)=DD*HW/(18.*L)+G*L/(30.*HW)
KL(1,24)=-G/20.
KL(1,28)=G*L/40.
KL(1,30)=-DD*4./9.*HW/L-G*L/(15.*HW)
KL(2,2)=DD*13./35.*L/HW+G*2./5.*HW/L
KL(2,6)=DD*11./210.*L**2/HW+G*HW/30.
KL(2,8)=G/20.
KL(2,9)=DD*9./70.*L/HW-G*2./5.*HW/L
KL(2,13)=-DD*13./420.*L**2/HW+G*HW/30.
KL(2,15)=G*2./5.
KL(2,16)=-G/20.
KL(2,17)=-DD*13./35.*L/HW+G*HW/(5.*L)
KL(2,21)=-DD*11./210.*L**2/HW+G*HW/60.
KL(2,23)=-G/20.
KL(2,24)=-DD*9./70.*L/HW-G*HW/(5.*L)
KL(2,28)=DD*13./420.*L**2/HW+G*HW/60.
KL(2,30)=-G*2./5.
KL(3,3)=DW*(156./35.*HW/L**3+156./35.*L/HW**3)
+         +2.*(1.-VNU)*DW*36./25./(HW*L)
+         +VNU*DW*2.*36./25./(HW*L)
KL(3,4)=DW*(22./35.*HW**2/L**3+78./35.*L/HW**2)
+         +2.*(1.-VNU)*DW*3./25./L
+         +VNU*DW*(33./25./L+3./25./L)
KL(3,5)=DW*(-78./35.*HW/L**2-22./35.*L**2/HW**3)
+         -2.*(1.-VNU)*DW*3./25./HW
+         +VNU*DW*(-3./25./HW-33./25./HW)
KL(3,7)=DW*(11./35.*HW**2/L**2+11./35.*L**2/HW**2)
+         +2.*(1.-VNU)*DW/100.
```

```
+        +VNU*DW*(11./100.+11./100.)
KL(3,10)=DW*(-156./35.*HW/L**3+54./35.*L/HW**3)
+        -2.*(1.-VNU)*DW*36./25./(HW*L)
+        +VNU*DW*(-36./25./(HW*L)-36./25./(HW*L))
KL(3,11)=DW*(-22./35.*HW**2/L**3+27./35.*L/HW**2)
+        -2.*(1.-VNU)*DW*3./25./L
+        +VNU*DW*(-33./25./L-3./25./L)
KL(3,12)=DW*(-78./35.*HW/L**2+13./35.*L**2/HW**3)
+        -2.*(1.-VNU)*DW*3./25./HW
+        +VNU*DW*(-3./25./HW-3./25./HW)
KL(3,14)=DW*(11./35.*HW**2/L**2-13./70.*L**2/HW**2)
+        +2.*(1.-VNU)*DW/100.
+        +VNU*DW*(11./100.+1./100.)
KL(3,18)=DW*(54./35.*HW/L**3-156./35.*L/HW**3)
+        -2.*(1.-VNU)*DW*36./25./(HW*L)
+        +VNU*DW*(-36./25./(HW*L)-36./25./(HW*L))
KL(3,19)=DW*(-13./35.*HW**2/L**3+78./35.*L/HW**2)
+        +2.*(1.-VNU)*DW*3./25./L
+        +VNU*DW*(3./25./L+3./25./L)
KL(3,20)=DW*(-27./35.*HW/L**2+22./35.*L**2/HW**3)
+        +2.*(1.-VNU)*DW*3./25./HW
+        +VNU*DW*(3./25./HW+33./25./HW)
KL(3,22)=DW*(-13./70.*HW**2/L**2+11./35.*L**2/HW**2)
+        +2.*(1.-VNU)*DW/100.
+        +VNU*DW*(1./100.+11./100.)
KL(3,25)=DW*(-54./35.*HW/L**3-54./35.*L/HW**3)
+        +2.*(1.-VNU)*DW*36./25./(HW*L)
+        +VNU*DW*(36./25./(HW*L)+36./25./(HW*L))
KL(3,26)=DW*(13./35.*HW**2/L**3+27./35.*L/HW**2)
+        -2.*(1.-VNU)*DW*3./25./L
+        +VNU*DW*(-3./25./L-3./25./L)
KL(3,27)=DW*(-27./35.*HW/L**2-13./35.*L**2/HW**3)
+        +2.*(1.-VNU)*DW*3./25./HW
+        +VNU*DW*(3./25./HW+3./25./HW)
KL(3,29)=DW*(-13./70.*HW**2/L**2-13./70.*L**2/HW**2)
+        +2.*(1.-VNU)*DW/100.
+        +VNU*DW*(1./100.+1./100.)
KL(4,4)=DW*(4./35.*HW**3/L**3+52./35.*L/HW)
+        +2.*(1.-VNU)*DW*4./25.*HW/L
+        +VNU*DW*2.*4./25.*HW/L
KL(4,5)=DW*(-11./35.*HW**2/L**2-11./35.*L**2/HW**2)
+        -2.*(1.-VNU)*DW/100.
```

+   +VNU*DW*(−121./100.−1./100.)
KL(4,7)=DW*(2./35.*HW**3/L**2+22./105.*L**2/HW)
+   +2.*(1.−VNU)*DW*HW/75.
+   +VNU*DW*(HW/75.+11./75.*HW)
KL(4,10)=DW*(−22./35.*HW**2/L**3+27./35.*L/HW**2)
+   −2.*(1.−VNU)*DW*3./25./L
+   +VNU*DW*(−33./25./L−3./25./L)
KL(4,11)=DW*(−4./35.*HW**3/L**3+18./35.*L/HW)
+   −2.*(1.−VNU)*DW*4./25.*HW/L
+   +VNU*DW*(−4./25.*HW/L−4./25.*HW/L)
KL(4,12)=DW*(−11./35.*HW**2/L**2+13./70.*L**2/HW**2)
+   −2.*(1.−VNU)*DW/100.
+   +VNU*DW*(−11./100.−1./100.)
KL(4,14)=DW*(2./35.*HW**3/L**2−13./105.*L**2/HW)
+   +2.*(1.−VNU)*DW*HW/75.
+   +VNU*DW*(HW/75.+HW/75.)
KL(4,18)=DW*(13./35.*HW**2/L**3−78./35.*L/HW**2)
+   −2.*(1.−VNU)*DW*3./25./L
+   +VNU*DW*(−3./25./L−3./25./L)
KL(4,19)=DW*(−3./35.*HW**3/L**3+26./35.*L/HW)
+   −2.*(1.−VNU)*DW*HW/(25.*L)
+   +VNU*DW*(−HW/(25.*L)−HW/(25.*L))
KL(4,20)=DW*(−13./70.*HW**2/L**2+11./35.*L**2/HW**2)
+   +2.*(1.−VNU)*DW/100.
+   +VNU*DW*(1./100.+11./100.)
KL(4,22)=DW*(−3./70.*HW**3/L**2+11./105.*L**2/HW)
+   −2.*(1.−VNU)*DW*HW/300.
+   +VNU*DW*(−HW/300.−11./300.*HW)
KL(4,25)=DW*(−13./35.*HW**2/L**3−27./35.*L/HW**2)
+   +2.*(1.−VNU)*DW*3./25./L
+   +VNU*DW*(3./25./L+3./25./L)
KL(4,26)=DW*(3./35.*HW**3/L**3+9./35.*L/HW)
+   +2.*(1.−VNU)*DW*HW/(25.*L)
+   +VNU*DW*(HW/(25.*L)+HW/(25.*L))
KL(4,27)=DW*(−13./70.*HW**2/L**2−13./70.*L**2/HW**2)
+   +2.*(1.−VNU)*DW/100.
+   +VNU*DW*(1./100.+1./100.)
KL(4,29)=DW*(−3./70.*HW**3/L**2−13./210.*L**2/HW)
+   −2.*(1.−VNU)*DW*HW/300.
+   +VNU*DW*(−HW/300.−HW/300.)
KL(5,5)=DW*(52./35.*HW/L+4./35.*L**3/HW**3)
+   +2.*(1.−VNU)*DW*4./25.*L/HW

```
+       +VNU*DW*2.*4./25.*L/HW
KL(5,7)=DW*(-22./105.*HW**2/L-2./35.*L**3/HW**2)
+       -2.*(1.-VNU)*DW*L/75.
+       +VNU*DW*(-11./75.*L-L/75.)
KL(5,10)=DW*(78./35.*HW/L**2-13./35.*L**2/HW**3)
+       +2.*(1.-VNU)*DW*3./25./HW
+       +VNU*DW*(3./25./HW+3./25./HW)
KL(5,11)=DW*(11./35.*HW**2/L**2-13./70.*L**2/HW**2)
+       +2.*(1.-VNU)*DW/100.
+       +VNU*DW*(11./100.+1./100.)
KL(5,12)=DW*(26./35.*HW/L-3./35.*L**3/HW**3)
+       -2.*(1.-VNU)*DW*L/(25.*HW)
+       +VNU*DW*(-L/(25.*HW)-L/(25.*HW))
KL(5,14)=DW*(-11./105.*HW**2/L+3./70.*L**3/HW**2)
+       +2.*(1.-VNU)*DW*L/300.
+       +VNU*DW*(11./300.*L+L/300.)
KL(5,18)=DW*(-27./35.*HW/L**2+22./35.*L**2/HW**3)
+       +2.*(1.-VNU)*DW*3./25./HW
+       +VNU*DW*(33./25./HW+3./25./HW)
KL(5,19)=DW*(13./70.*HW**2/L**2-11./35.*L**2/HW**2)
+       -2.*(1.-VNU)*DW/100.
+       +VNU*DW*(-11./100.-1./100.)
KL(5,20)=DW*(18./35.*HW/L-4./35.*L**3/HW**3)
+       -2.*(1.-VNU)*DW*4./25.*L/HW
+       +VNU*DW*(-4./25.*L/HW-4./25.*L/HW)
KL(5,22)=DW*(13./105.*HW**2/L-2./35.*L**3/HW**2)
+       -2.*(1.-VNU)*DW*L/75.
+       +VNU*DW*(-L/75.-L/75.)
KL(5,25)=DW*(27./35.*HW/L**2+13./35.*L**2/HW**3)
+       -2.*(1.-VNU)*DW*3./25./HW
+       +VNU*DW*(-3./25./HW-3./25./HW)
KL(5,26)=DW*(-13./70.*HW**2/L**2-13./70.*L**2/HW**2)
+       +2.*(1.-VNU)*DW/100.
+       +VNU*DW*(1./100.+1./100.)
KL(5,27)=DW*(9./35.*HW/L+3./35.*L**3/HW**3)
+       +2.*(1.-VNU)*DW*L/(25.*HW)
+       +VNU*DW*(L/(25.*HW)+L/(25.*HW))
KL(5,29)=DW*(13./210.*HW**2/L+3./70.*L**3/HW**2)
+       +2.*(1.-VNU)*DW*L/300.
+       +VNU*DW*(L/300.+L/300.)
KL(6,6)=DD*L**3/(105.*HW)+G*2./45.*HW*L
KL(6,8)=G*L/40.
```

KL(6,9)=DD*13./420.*L**2/HW–G*HW/30.
KL(6,13)=–DD*L**3/(140.*HW)–G*HW*L/90.
KL(6,15)=G*L/30.
KL(6,16)=G*7./120.*L
KL(6,17)=–DD*11./210.*L**2/HW+G*HW/60.
KL(6,21)=–DD*L**3/(105.*HW)+G*HW*L/45.
KL(6,23)=–G*L/40.
KL(6,24)=–DD*13./420.*L**2/HW–G*HW/60.
KL(6,28)=DD*L**3/(140.*HW)–G*HW*L/180.
KL(6,30)=–G*L/30.
KL(7,7)=DW*(4./105.*HW**3/L+4./105.*L**3/HW)
+   +2.*(1.–VNU)*DW*4./225.*HW*L
+   +VNU*DW*2.*4./225.*HW*L
KL(7,10)=DW*(–11./35.*HW**2/L**2+13./70.*L**2/HW**2)
+   –2.*(1.–VNU)*DW/100.
+   +VNU*DW*(–11./100.–1./100.)
KL(7,11)=DW*(–2./35.*HW**3/L**2+13./105.*L**2/HW)
+   –2.*(1.–VNU)*DW*HW/75.
+   +VNU*DW*(–HW/75.–HW/75.)
KL(7,12)=DW*(–11./105.*HW**2/L+3./70.*L**3/HW**2)
+   +2.*(1.–VNU)*DW*L/300.
+   +VNU*DW*(11./300.*L+L/300.)
KL(7,14)=DW*(2./105.*HW**3/L–L**3/(35.*HW))
+   –2.*(1.–VNU)*DW*HW*L/225.
+   +VNU*DW*(–HW*L/225.–HW*L/225.)
KL(7,18)=DW*(13./70.*HW**2/L**2–11./35.*L**2/HW**2)
+   –2.*(1.–VNU)*DW/100.
+   +VNU*DW*(–11./100.–1./100.)
KL(7,19)=DW*(–3./70.*HW**3/L**2+11./105.*L**2/HW)
+   –2.*(1.–VNU)*DW*HW/300.
+   +VNU*DW*(–11./300.*HW–HW/300.)
KL(7,20)=DW*(–13./105.*HW**2/L+2./35.*L**3/HW**2)
+   +2.*(1.–VNU)*DW*L/75.
+   +VNU*DW*(L/75.+L/75.)
KL(7,22)=DW*(–HW**3/(35.*L)+2./105.*L**3/HW)
+   –2.*(1.–VNU)*DW*HW*L/225.
+   +VNU*DW*(–HW*L/225.–HW*L/225.)
KL(7,25)=DW*(–13./70.*HW**2/L**2–13./70.*L**2/HW**2)
+   +2.*(1.–VNU)*DW/100.
+   +VNU*DW*(1./100.+1./100.)
KL(7,26)=DW*(3./70.*HW**3/L**2+13./210.*L**2/HW)
+   +2.*(1.–VNU)*DW*HW/300.

```
+        +VNU*DW*(HW/300.+HW/300.)
KL(7,27)=DW*(-13./210.*HW**2/L-3./70.*L**3/HW**2)
+        -2.*(1.-VNU)*DW*L/300.
+        +VNU*DW*(-L/300.-L/300.)
KL(7,29)=DW*(-HW**3/(70.*L)-L**3/(70.*HW))
+        +2.*(1.-VNU)*DW*HW*L/900.
+        +VNU*DW*(HW*L/900.+HW*L/900.)
KL(8,8)=DD*7./9.*HW/L+G*2./15.*L/HW
KL(8,9)=-G/20.
KL(8,13)=-G*7./120.*L
KL(8,15)=-DD*8./9.*HW/L+G*L/(15.*HW)
KL(8,16)=DD*HW/(18.*L)+G*L/(30.*HW)
KL(8,17)=G/20.
KL(8,21)=G*L/40.
KL(8,23)=DD*7./18.*HW/L-G*2./15.*L/HW
KL(8,24)=-G/20.
KL(8,28)=-G*7./120.*L
KL(8,30)=-DD*4./9.*HW/L-G*L/(15.*HW)
KL(9,9)=DD*13./35.*L/HW+G*2./5.*HW/L
KL(9,13)=-DD*11./210.*L**2/HW-G*HW/30.
KL(9,15)=-G*2./5.
KL(9,16)=G/20.
KL(9,17)=-DD*9./70.*L/HW-G*HW/(5.*L)
KL(9,21)=-DD*13./420.*L**2/HW-G*HW/60.
KL(9,23)=G/20.
KL(9,24)=-DD*13./35.*L/HW+G*HW/(5.*L)
KL(9,28)=DD*11./210.*L**2/HW-G*HW/60.
KL(9,30)=G*2./5.
KL(10,10)=DW*(156./35.*HW/L**3+156./35.*L/HW**3)
+        +2.*(1.-VNU)*DW*36./25./(HW*L)
+        +VNU*DW*2.*36./25./(HW*L)
KL(10,11)=DW*(22./35.*HW**2/L**3+78./35.*L/HW**2)
+        +2.*(1.-VNU)*DW*3./25./L
+        +VNU*DW*(33./25./L+3./25./L)
KL(10,12)=DW*(78./35.*HW/L**2+22./35.*L**2/HW**3)
+        +2.*(1.-VNU)*DW*3./25./HW
+        +VNU*DW*(3./25./HW+33./25./HW)
KL(10,14)=DW*(-11./35.*HW**2/L**2-11./35.*L**2/HW**2)
+        -2.*(1.-VNU)*DW/100.
+        +VNU*DW*(-11./100.-11./100.)
KL(10,18)=DW*(-54./35.*HW/L**3-54./35.*L/HW**3)
+        +2.*(1.-VNU)*DW*36./25./(HW*L)
```

$$+ \qquad +VNU*DW*(36./25./(HW*L)+36./25./(HW*L))$$
$$KL(10,19)=DW*(13./35.*HW**2/L**3+27./35.*L/HW**2)$$
$$+ \qquad -2.*(1.-VNU)*DW*3./25./L$$
$$+ \qquad +VNU*DW*(-3./25./L-3./25./L)$$
$$KL(10,20)=DW*(27./35.*HW/L**2+13./35.*L**2/HW**3)$$
$$+ \qquad -2.*(1.-VNU)*DW*3./25./HW$$
$$+ \qquad +VNU*DW*(-3./25./HW-3./25./HW)$$
$$KL(10,22)=DW*(13./70.*HW**2/L**2+13./70.*L**2/HW**2)$$
$$+ \qquad -2.*(1.-VNU)*DW/100.$$
$$+ \qquad +VNU*DW*(-1./100.-1./100.)$$
$$KL(10,25)=DW*(54./35.*HW/L**3-156./35.*L/HW**3)$$
$$+ \qquad -2.*(1.-VNU)*DW*36./25./(HW*L)$$
$$+ \qquad +VNU*DW*(-36./25./(HW*L)-36./25./(HW*L))$$
$$KL(10,26)=DW*(-13./35.*HW**2/L**3+78./35.*L/HW**2)$$
$$+ \qquad +2.*(1.-VNU)*DW*3./25./L$$
$$+ \qquad +VNU*DW*(3./25./L+3./25./L)$$
$$KL(10,27)=DW*(27./35.*HW/L**2-22./35.*L**2/HW**3)$$
$$+ \qquad -2.*(1.-VNU)*DW*3./25./HW$$
$$+ \qquad +VNU*DW*(-3./25./HW-33./25./HW)$$
$$KL(10,29)=DW*(13./70.*HW**2/L**2-11./35.*L**2/HW**2)$$
$$+ \qquad -2.*(1.-VNU)*DW/100.$$
$$+ \qquad +VNU*DW*(-1./100.-11./100.)$$
$$KL(11,11)=DW*(4./35.*HW**3/L**3+52./35.*L/HW)$$
$$+ \qquad +2.*(1.-VNU)*DW*4./25.*HW/L$$
$$+ \qquad +VNU*DW*2.*4./25.*HW/L$$
$$KL(11,12)=DW*(11./35.*HW**2/L**2+11./35.*L**2/HW**2)$$
$$+ \qquad +2.*(1.-VNU)*DW/100.$$
$$+ \qquad +VNU*DW*(121./100.+1./100.)$$
$$KL(11,14)=DW*(-2./35.*HW**3/L**2-22./105.*L**2/HW)$$
$$+ \qquad -2.*(1.-VNU)*DW*HW/75.$$
$$+ \qquad +VNU*DW*(-HW/75.-11./75.*HW)$$
$$KL(11,18)=DW*(-13./35.*HW**2/L**3-27./35.*L/HW**2)$$
$$+ \qquad +2.*(1.-VNU)*DW*3./25./L$$
$$+ \qquad +VNU*DW*(3./25./L+3./25./L)$$
$$KL(11,19)=DW*(3./35.*HW**3/L**3+9./35.*L/HW)$$
$$+ \qquad +2.*(1.-VNU)*DW*HW/(25.*L)$$
$$+ \qquad +VNU*DW*(HW/(25.*L)+HW/(25.*L))$$
$$KL(11,20)=DW*(13./70.*HW**2/L**2+13./70.*L**2/HW**2)$$
$$+ \qquad -2.*(1.-VNU)*DW/100.$$
$$+ \qquad +VNU*DW*(-1./100.-1./100.)$$
$$KL(11,22)=DW*(3./70.*HW**3/L**2+13./210.*L**2/HW)$$
$$+ \qquad +2.*(1.-VNU)*DW*HW/300.$$

```
+        +VNU*DW*(HW/300.+HW/300.)
KL(11,25)=DW*(13./35.*HW**2/L**3-78./35.*L/HW**2)
+        -2.*(1.-VNU)*DW*3./25./L
+        +VNU*DW*(-3./25./L-3./25./L)
KL(11,26)=DW*(-3./35.*HW**3/L**3+26./35.*L/HW)
+        -2.*(1.-VNU)*DW*HW/(25.*L)
+        +VNU*DW*(-HW/(25.*L)-HW/(25.*L))
KL(11,27)=DW*(13./70.*HW**2/L**2-11./35.*L**2/HW**2)
+        -2.*(1.-VNU)*DW/100.
+        +VNU*DW*(-1./100.-11./100.)
KL(11,29)=DW*(3./70.*HW**3/L**2-11./105.*L**2/HW)
+        +2.*(1.-VNU)*DW*HW/300.
+        +VNU*DW*(HW/300.+11./300.*HW)
KL(12,12)=DW*(52./35.*HW/L+4./35.*L**3/HW**3)
+        +2.*(1.-VNU)*DW*4./25.*L/HW
+        +VNU*DW*2.*4./25.*L/HW
KL(12,14)=DW*(-22./105.*HW**2/L-2./35.*L**3/HW**2)
+        -2.*(1.-VNU)*DW*L/75.
+        +VNU*DW*(-11./75.*L-L/75.)
KL(12,18)=DW*(-27./35.*HW/L**2-13./35.*L**2/HW**3)
+        +2.*(1.-VNU)*DW*3./25./HW
+        +VNU*DW*(3./25./HW+3./25./HW)
KL(12,19)=DW*(13./70.*HW**2/L**2+13./70.*L**2/HW**2)
+        -2.*(1.-VNU)*DW/100.
+        +VNU*DW*(-1./100.-1./100.)
KL(12,20)=DW*(9./35.*HW/L+3./35.*L**3/HW**3)
+        +2.*(1.-VNU)*DW*L/(25.*HW)
+        +VNU*DW*(L/(25.*HW)+L/(25.*HW))
KL(12,22)=DW*(13./210.*HW**2/L+3./70.*L**3/HW**2)
+        +2.*(1.-VNU)*DW*L/300.
+        +VNU*DW*(L/300.+L/300.)
KL(12,25)=DW*(27./35.*HW/L**2-22./35.*L**2/HW**3)
+        -2.*(1.-VNU)*DW*3./25./HW
+        +VNU*DW*(-33./25./HW-3./25./HW)
KL(12,26)=DW*(-13./70.*HW**2/L**2+11./35.*L**2/HW**2)
+        +2.*(1.-VNU)*DW/100.
+        +VNU*DW*(11./100.+1./100.)
KL(12,27)=DW*(18./35.*HW/L-4./35.*L**3/HW**3)
+        -2.*(1.-VNU)*DW*4./25.*L/HW
+        +VNU*DW*(-4./25.*L/HW-4./25.*L/HW)
KL(12,29)=DW*(13./105.*HW**2/L-2./35.*L**3/HW**2)
+        -2.*(1.-VNU)*DW*L/75.
```

```
+       +VNU*DW*(-L/75.-L/75.)
KL(13,13)=DD*L**3/(105.*HW)+G*2./45.*HW*L
KL(13,15)=G*L/30.
KL(13,16)=-G*L/40.
KL(13,17)=DD*13./420.*L**2/HW+G*HW/60.
KL(13,21)=DD*L**3/(140.*HW)-G*HW*L/180.
KL(13,23)=G*7./120.*L
KL(13,24)=DD*11./210.*L**2/HW-G*HW/60.
KL(13,28)=-DD*L**3/(105.*HW)+G*HW*L/45.
KL(13,30)=-G*L/30.
KL(14,14)=DW*(4./105.*HW**3/L+4./105.*L**3/HW)
+       +2.*(1.-VNU)*DW*4./225.*HW*L
+       +VNU*DW*2.*4./225.*HW*L
KL(14,18)=DW*(13./70.*HW**2/L**2+13./70.*L**2/HW**2)
+       -2.*(1.-VNU)*DW/100.
+       +VNU*DW*(-1./100.-1./100.)
KL(14,19)=DW*(-3./70.*HW**3/L**2-13./210.*L**2/HW)
+       -2.*(1.-VNU)*DW*HW/300.
+       +VNU*DW*(-HW/300.-HW/300.)
KL(14,20)=DW*(-13./210.*HW**2/L-3./70.*L**3/HW**2)
+       -2.*(1.-VNU)*DW*L/300.
+       +VNU*DW*(-L/300.-L/300.)
KL(14,22)=DW*(-HW**3/(70.*L)-L**3/(70.*HW))
+       +2.*(1.-VNU)*DW*HW*L/900.
+       +VNU*DW*(HW*L/900.+HW*L/900.)
KL(14,25)=DW*(-13./70.*HW**2/L**2+11./35.*L**2/HW**2)
+       +2.*(1.-VNU)*DW/100.
+       +VNU*DW*(11./100.+1./100.)
KL(14,26)=DW*(3./70.*HW**3/L**2-11./105.*L**2/HW)
+       +2.*(1.-VNU)*DW*HW/300.
+       +VNU*DW*(11./300.*HW+HW/300.)
KL(14,27)=DW*(-13./105.*HW**2/L+2./35.*L**3/HW**2)
+       +2.*(1.-VNU)*DW*L/75.
+       +VNU*DW*(L/75.+L/75.)
KL(14,29)=DW*(-HW**3/(35.*L)+2./105.*L**3/HW)
+       -2.*(1.-VNU)*DW*HW*L/225.
+       +VNU*DW*(-HW*L/225.-HW*L/225.)
KL(15,15)=DD*16./9.*HW/L+G*8./15.*L/HW
KL(15,16)=-DD*4./9.*HW/L-G*L/(15.*HW)
KL(15,17)=G*2./5.
KL(15,21)=G*L/30.
KL(15,23)=-DD*4./9.*HW/L-G*L/(15.*HW)
```

KL(15,24)=–G*2./5.
KL(15,28)=G*L/30.
KL(15,30)=DD*8./9.*HW/L–G*8./15.*L/HW
KL(16,16)=DD*7./9.*HW/L+G*2./15.*L/HW
KL(16,17)=–G/20.
KL(16,21)=G*7./120.*L
KL(16,23)=DD*HW/(9.*L)–G*L/(30.*HW)
KL(16,24)=G/20.
KL(16,28)=–G*L/40.
KL(16,30)=–DD*8./9.*HW/L+G*L/(15.*HW)
KL(17,17)=DD*13./35.*L/HW+G*2./5.*HW/L
KL(17,21)=DD*11./210.*L**2/HW+G*HW/30.
KL(17,23)=–G/20.
KL(17,24)=DD*9./70.*L/HW–G*2./5.*HW/L
KL(17,28)=–DD*13./420.*L**2/HW+G*HW/30.
KL(17,30)=–G*2./5.
KL(18,18)=DW*(156./35.*HW/L**3+156./35.*L/HW**3)
+   +2.*(1.–VNU)*DW*36./25./(HW*L)
+   +VNU*DW*2.*36./25./(HW*L)
KL(18,19)=DW*(–22./35.*HW**2/L**3–78./35.*L/HW**2)
+   –2.*(1.–VNU)*DW*3./25./L
+   +VNU*DW*(–33./25./L–3./25./L)
KL(18,20)=DW*(–78./35.*HW/L**2–22./35.*L**2/HW**3)
+   –2.*(1.–VNU)*DW*3./25./HW
+   +VNU*DW*(–3./25./HW–33./25./HW)
KL(18,22)=DW*(–11./35.*HW**2/L**2–11./35.*L**2/HW**2)
+   –2.*(1.–VNU)*DW/100.
+   +VNU*DW*(–11./100.–11./100.)
KL(18,25)=DW*(–156./35.*HW/L**3+54./35.*L/HW**3)
+   –2.*(1.–VNU)*DW*36./25./(HW*L)
+   +VNU*DW*(–36./25./(HW*L)–36./25./(HW*L))
KL(18,26)=DW*(22./35.*HW**2/L**3–27./35.*L/HW**2)
+   +2.*(1.–VNU)*DW*3./25./L
+   +VNU*DW*(33./25./L+3./25./L)
KL(18,27)=DW*(–78./35.*HW/L**2+13./35.*L**2/HW**3)
+   –2.*(1.–VNU)*DW*3./25./HW
+   +VNU*DW*(–3./25./HW–3./25./HW)
KL(18,29)=DW*(–11./35.*HW**2/L**2+13./70.*L**2/HW**2)
+   –2.*(1.–VNU)*DW/100.
+   +VNU*DW*(–11./100.–1./100.)
KL(19,19)=DW*(4./35.*HW**3/L**3+52./35.*L/HW)
+   +2.*(1.–VNU)*DW*4./25.*HW/L

```
+        +VNU*DW*2.*4./25.*HW/L
KL(19,20)=DW*(11./35.*HW**2/L**2+11./35.*L**2/HW**2)
+        +2.*(1.-VNU)*DW/100.
+        +VNU*DW*(121./100.+1./100.)
KL(19,22)=DW*(2./35.*HW**3/L**2+22./105.*L**2/HW)
+        +2.*(1.-VNU)*DW*HW/75.
+        +VNU*DW*(HW/75.+11./75.*HW)
KL(19,25)=DW*(22./35.*HW**2/L**3-27./35.*L/HW**2)
+        +2.*(1.-VNU)*DW*3./25./L
+        +VNU*DW*(33./25./L+3./25./L)
KL(19,26)=DW*(-4./35.*HW**3/L**3+18./35.*L/HW)
+        -2.*(1.-VNU)*DW*4./25.*HW/L
+        +VNU*DW*(-4./25.*HW/L-4./25.*HW/L)
KL(19,27)=DW*(11./35.*HW**2/L**2-13./70.*L**2/HW**2)
+        +2.*(1.-VNU)*DW/100.
+        +VNU*DW*(11./100.+1./100.)
KL(19,29)=DW*(2./35.*HW**3/L**2-13./105.*L**2/HW)
+        +2.*(1.-VNU)*DW*HW/75.
+        +VNU*DW*(HW/75.+HW/75.)
KL(20,20)=DW*(52./35.*HW/L+4./35.*L**3/HW**3)
+        +2.*(1.-VNU)*DW*4./25.*L/HW
+        +VNU*DW*2.*4./25.*L/HW
KL(20,22)=DW*(22./105.*HW**2/L+2./35.*L**3/HW**2)
+        +2.*(1.-VNU)*DW*L/75.
+        +VNU*DW*(11./75.*L+L/75.)
KL(20,25)=DW*(78./35.*HW/L**2-13./35.*L**2/HW**3)
+        +2.*(1.-VNU)*DW*3./25./HW
+        +VNU*DW*(3./25./HW+3./25./HW)
KL(20,26)=DW*(-11./35.*HW**2/L**2+13./70.*L**2/HW**2)
+        -2.*(1.-VNU)*DW/100.
+        +VNU*DW*(-11./100.-1./100.)
KL(20,27)=DW*(26./35.*HW/L-3./35.*L**3/HW**3)
+        -2.*(1.-VNU)*DW*L/(25.*HW)
+        +VNU*DW*(-L/(25.*HW)-L/(25.*HW))
KL(20,29)=DW*(11./105.*HW**2/L-3./70.*L**3/HW**2)
+        -2.*(1.-VNU)*DW*L/300.
+        +VNU*DW*(-11./300.*L-L/300.)
KL(21,21)=DD*L**3/(105.*HW)+G*2./45.*HW*L
KL(21,23)=-G*L/40.
KL(21,24)=DD*13./420.*L**2/HW-G*HW/30.
KL(21,28)=-DD*L**3/(140.*HW)-G*HW*L/90.
KL(21,30)=-G*L/30.
```

KL(22,22)=DW*(4./105.*HW**3/L+4./105.*L**3/HW)
+   +2.*(1.–VNU)*DW*4./225.*HW*L
+   +VNU*DW*2.*4./225.*HW*L
KL(22,25)=DW*(11./35.*HW**2/L**2–13./70.*L**2/HW**2)
+   +2.*(1.–VNU)*DW/100.
+   +VNU*DW*(11./100.+1./100.)
KL(22,26)=DW*(–2./35.*HW**3/L**2+13./105.*L**2/HW)
+   –2.*(1.–VNU)*DW*HW/75.
+   +VNU*DW*(–HW/75.–HW/75.)
KL(22,27)=DW*(11./105.*HW**2/L–3./70.*L**3/HW**2)
+   –2.*(1.–VNU)*DW*L/300.
+   +VNU*DW*(–11./300.*L–L/300.)
KL(22,29)=DW*(2./105.*HW**3/L–L**3/(35.*HW))
+   –2.*(1.–VNU)*DW*HW*L/225.
+   +VNU*DW*(–HW*L/225.–HW*L/225.)
KL(23,23)=DD*7./9.*HW/L+G*2./15.*L/HW
KL(23,24)=G/20.
KL(23,28)=G*7./120.*L
KL(23,30)=–DD*8./9.*HW/L+G*L/(15.*HW)
KL(24,24)=DD*13./35.*L/HW+G*2./5.*HW/L
KL(24,28)=–DD*11./210.*L**2/HW–G*HW/30.
KL(24,30)=G*2./5.
KL(25,25)=DW*(156./35.*HW/L**3+156./35.*L/HW**3)
+   +2.*(1.–VNU)*DW*36./25./(HW*L)
+   +VNU*DW*2.*36./25./(HW*L)
KL(25,26)=DW*(–22./35.*HW**2/L**3–78./35.*L/HW**2)
+   –2.*(1.–VNU)*DW*3./25./L
+   +VNU*DW*(–33./25./L–3./25./L)
KL(25,27)=DW*(78./35.*HW/L**2+22./35.*L**2/HW**3)
+   +2.*(1.–VNU)*DW*3./25./HW
+   +VNU*DW*(3./25./HW+33./25./HW)
KL(25,29)=DW*(11./35.*HW**2/L**2+11./35.*L**2/HW**2)
+   +2.*(1.–VNU)*DW/100.
+   +VNU*DW*(11./100.+11./100.)
KL(26,26)=DW*(4./35.*HW**3/L**3+52./35.*L/HW)
+   +2.*(1.–VNU)*DW*4./25.*HW/L
+   +VNU*DW*2.*4./25.*HW/L
KL(26,27)=DW*(–11./35.*HW**2/L**2–11./35.*L**2/HW**2)
+   –2.*(1.–VNU)*DW/100.
+   +VNU*DW*(–121./100.–1./100.)
KL(26,29)=DW*(–2./35.*HW**3/L**2–22./105.*L**2/HW)
+   –2.*(1.–VNU)*DW*HW/75.

```
+        +VNU*DW*(-HW/75.-11./75.*HW)
KL(27,27)=DW*(52./35.*HW/L+4./35.*L**3/HW**3)
+        +2.*(1.-VNU)*DW*4./25.*L/HW
+        +VNU*DW*2.*4./25.*L/HW
KL(27,29)=DW*(22./105.*HW**2/L+2./35.*L**3/HW**2)
+        +2.*(1.-VNU)*DW*L/75.
+        +VNU*DW*(11./75.*L+L/75.)
KL(28,28)=DD*L**3/(105.*HW)+G*2./45.*HW*L
KL(28,30)=-G*L/30.
KL(29,29)=DW*(4./105.*HW**3/L+4./105.*L**3/HW)
+        +2.*(1.-VNU)*DW*4./225.*HW*L
+        +VNU*DW*2.*4./225.*HW*L
KL(30,30)=DD*16./9.*HW/L+G*8./15.*L/HW
```

The upper triangle of the geometric stiffness matrix for the thin plate element **k**$_G$:

```
KG(1,1)=-(55.*FXB1-15.*FXB2-11.*FXT1+3.*FXT2)/(36.*L)
$        +2.*(FYT1+FYT2)/(15.*HW)
$        +2.*(FYB2+FYT2)/(4.*HW)
KG(1,8)=-(5.*FXB1-5.*FXB2-FXT1+FXT2)/(36.*L)
$        +(-FYT1-FYT2)/(30.*HW)
KG(1,15)=(15.*FXB1-5.*FXB2-3.*FXT1+FXT2)/(9.*L)
$        +(FYT1+FYT2)/(15.*HW)
KG(1,16)=-(11.*FXB1-3.*FXB2+11.*FXT1-3.*FXT2)/(36.*L)
$        +2.*(-FYT1-FYT2)/(15.*HW)
KG(1,23)=-(FXB1-FXB2+FXT1-FXT2)/(36.*L)
$        +(FYT1+FYT2)/(30.*HW)
$        +2.*(FYB2+FYT2)/(12.*HW)
KG(1,30)=(3.*FXB1-FXB2+3.*FXT1-FXT2)/(9.*L)
$        +(-FYT1-FYT2)/(15.*HW)
$        +2.*(-FYB2-FYT2)/(3.*HW)
KG(2,2)=-(5.*FXB1-5.*FXB2-FXT1+FXT2)/(10.*L)
$        +13.*(FYT1+FYT2)/(35.*HW)
$        +2.*(FYB2+FYT2)/(4.*HW)
KG(2,6)=(5.*FXB2-FXT2)/60.
$        +11.*L*(FYT1+FYT2)/(210.*HW)
KG(2,9)=(5.*FXB1-5.*FXB2-FXT1+FXT2)/(10.*L)
$        +9.*(FYT1+FYT2)/(70.*HW)
KG(2,13)=-(5.*FXB1-FXT1)/60.
$        +13.*L*(-FYT1-FYT2)/(420.*HW)
KG(2,17)=-(FXB1-FXB2+FXT1-FXT2)/(10.*L)
```

$ +13.*(–FYT1–FYT2)/(35.*HW)
KG(2,21)=(FXB2+FXT2)/60.
$ +11.*L*(–FYT1–FYT2)/(210.*HW)
$ +2.*(–L*FYB2–L*FYT2)/(20.*HW)
KG(2,24)=(FXB1–FXB2+FXT1–FXT2)/(10.*L)
$ +9.*(–FYT1–FYT2)/(70.*HW)
$ +2.*(–FYB2–FYT2)/(4.*HW)
KG(2,28)=–(FXB1+FXT1)/60.
$ +13.*L*(FYT1+FYT2)/(420.*HW)
$ +2.*(L*FYB2+L*FYT2)/(20.*HW)
KG(3,3)=–6.*(17.*FXB1–17.*FXB2–4.*(FXT1–FXT2))/(175.*L)
$ +78.*(FYT1+FYT2)/(175.*HW)
$ +(FYB2+FYT2)/(2.*HW)
KG(3,4)=–HW*(23.*FXB1–23.*FXB2–FXT1+FXT2)/(350.*L)
$ +13.*(FYT1+FYT2)/350.
KG(3,5)=–(17.*FXB2–4.*FXT2)/175.
$ +11.*L*(–FYT1–FYT2)/(175.*HW)
KG(3,7)=HW*(23.*FXB2–FXT2)/2100.
$ +11.*L*(FYT1+FYT2)/2100.
$ +(–L*FYB2–L*FYT2)/50.
KG(3,10)=6.*(17.*FXB1–17.*FXB2–4.*(FXT1–FXT2))/(175.*L)
$ +27.*(FYT1+FYT2)/(175.*HW)
KG(3,11)=HW*(23.*FXB1–23.*FXB2–FXT1+FXT2)/(350.*L)
$ +9.*(FYT1+FYT2)/700.
$ +(–FYB2–FYT2)/10.
KG(3,12)=(17.*FXB1–4.*FXT1)/175.
$ +13.*L*(FYT1+FYT2)/(350.*HW)
KG(3,14)=–HW*(23.*FXB1–FXT1)/2100.
$ +13.*L*(–FYT1–FYT2)/4200.
$ +(L*FYB2+L*FYT2)/50.
KG(3,18)=–27.*(FXB1–FXB2+FXT1–FXT2)/(350.*L)
$ +78.*(–FYT1–FYT2)/(175.*HW)
KG(3,19)=HW*(8.*FXB1–8.*FXB2+5.*(FXT1–FXT2))/(350.*L)
$ +13.*(FYT1+FYT2)/350.
KG(3,20)=–9.*(FXB2+FXT2)/700.
$ +11.*L*(FYT1+FYT2)/(175.*HW)
$ +(L*FYB2+L*FYT2)/(10.*HW)
KG(3,22)=–HW*(8.*FXB2+5.*FXT2)/2100.
$ +11.*L*(FYT1+FYT2)/2100.
$ +(L*FYB2+L*FYT2)/50.
KG(3,25)=27.*(FXB1–FXB2+FXT1–FXT2)/(350.*L)
$ +27.*(–FYT1–FYT2)/(175.*HW)

```
$        +(-FYB2-FYT2)/(2.*HW)
KG(3,26)=-HW*(8.*FXB1-8.*FXB2+5.*(FXT1-FXT2))/(350.*L)
$        +9.*(FYT1+FYT2)/700.
$        +(FYB2+FYT2)/10.
KG(3,27)=9.*(FXB1+FXT1)/700.
$        +13.*L*(-FYT1-FYT2)/(350.*HW)
$        +(-L*FYB2-L*FYT2)/(10.*HW)
KG(3,29)=HW*(8.*FXB1+5.*FXT1)/2100.
$        +13.*L*(-FYT1-FYT2)/4200.
$        +L*(-FYB2-FYT2)/50.
KG(4,4)=-HW**2*(7.*FXB1-7.*FXB2+FXT1-FXT2)/(700.*L)
$        +26.*HW*(FYT1+FYT2)/525.
KG(4,5)=-HW*(23.*FXB2-FXT2)/2100.
$        +11.*L*(-FYT1-FYT2)/2100.
$        +L*(-FYB2-FYT2)/50.
KG(4,7)=HW**2*(7.*FXB2+FXT2)/4200.
$        +11.*HW*L*(FYT1+FYT2)/1575.
KG(4,10)=HW*(23.*FXB1-23.*FXB2-FXT1+FXT2)/(350.*L)
$        +9.*(FYT1+FYT2)/700.
$        +(FYB2+FYT2)/10.
KG(4,11)=HW**2*(7.*FXB1-7.*FXB2+FXT1-FXT2)/(700.*L)
$        +3.*HW*(FYT1+FYT2)/175.
KG(4,12)=HW*(23.*FXB1-FXT1)/2100.
$        +13.*L*(FYT1+FYT2)/4200.
$        +L*(FYB2+FYT2)/50.
KG(4,14)=-HW**2*(7.*FXB1+FXT1)/4200.
$        +13.*HW*L*(-FYT1-FYT2)/3150.
KG(4,18)=-HW*(5.*FXB1-5.*FXB2+8.*(FXT1-FXT2))/(350.*L)
$        +(-13.*FYT1-13.*FYT2)/350.
KG(4,19)=3.*HW**2*(FXB1-FXB2+FXT1-FXT2)/(700.*L)
$        +13.*HW*(-FYT1-FYT2)/1050.
KG(4,20)=-HW*(5.*FXB2+8.*FXT2)/2100.
$        +11.*L*(FYT1+FYT2)/2100.
$        +L*(FYB2+FYT2)/50.
KG(4,22)=-HW**2*(FXB2+FXT2)/1400.
$        +11.*HW*L*(-FYT1-FYT2)/6300.
$        +HW*L*(FYB2+FYT2)/300.
KG(4,25)=HW*(5.*FXB1-5.*FXB2+8.*(FXT1-FXT2))/(350.*L)
$        +(-9.*FYT1-9.*FYT2)/700.
$        +(-FYB2-FYT2)/10.
KG(4,26)=-3.*HW**2*(FXB1-FXB2+FXT1-FXT2)/(700.*L)
$        +3.*HW*(-FYT1-FYT2)/700.
```

```
$       +HW*(FYB2+FYT2)/60.
KG(4,27)=HW*(5.*FXB1+8.*FXT1)/2100.
$       +13.*L*(–FYT1–FYT2)/4200.
$       +L*(–FYB2–FYT2)/50.
KG(4,29)=HW**2*(FXB1+FXT1)/1400.
$       +13.*HW*L*(FYT1+FYT2)/12600.
$       +HW*L*(–FYB2–FYT2)/300.
KG(5,5)=–L*(51.*FXB1–17.*FXB2–4.*(3.*FXT1–FXT2))/525.
$       +2.*L**2*(FYT1+FYT2)/(175.*HW)
KG(5,7)=HW*L*(69.*FXB1–23.*FXB2–3.*FXT1+FXT2)/6300.
$       +L**2*(–FYT1–FYT2)/1050.
KG(5,10)=(17.*FXB2–4.*FXT2)/175.
$       +13.*L*(–FYT1–FYT2)/(350.*HW)
KG(5,11)=HW*(23.*FXB2–FXT2)/2100.
$       +13.*L*(–FYT1–FYT2)/4200.
$       +L*(FYB2+FYT2)/50.
KG(5,12)=L*(17.*FXB1–17.*FXB2–4.*(FXT1–FXT2))/1050.
$       +3.*L**2*(–FYT1–FYT2)/(350.*HW)
KG(5,14)=–HW*L*(23.*FXB1–23.*FXB2–FXT1+FXT2)/12600.
$       +L**2*(FYT1+FYT2)/1400.
$       +L**2*(–FYB2–FYT2)/300.
KG(5,18)=–9.*(FXB2+FXT2)/700.
$       +11.*L*(FYT1+FYT2)/(175.*HW)
$       +L*(–FYB2–FYT2)/(10.*HW)
KG(5,19)=HW*(8.*FXB2+5.*FXT2)/2100.
$       +11.*L*(–FYT1–FYT2)/2100.
$       +L*(FYB2+FYT2)/50.
KG(5,20)=–3.*L*(3.*FXB1–FXB2+3.*FXT1–FXT2)/700.
$       +2.*L**2*(–FYT1–FYT2)/(175.*HW)
KG(5,22)=–HW*L*(24.*FXB1–8.*FXB2+5.*(3.*FXT1–FXT2))/6300.
$       +L**2*(–FYT1–FYT2)/1050.
KG(5,25)=9.*(FXB2+FXT2)/700.
$       +13.*L*(FYT1+FYT2)/(350.*HW)
$       +L*(FYB2+FYT2)/(10.*HW)
KG(5,26)=–HW*(8.*FXB2+5.*FXT2)/2100.
$       +13.*L*(–FYT1–FYT2)/4200.
$       +L*(–FYB2–FYT2)/50.
KG(5,27)=3.*L*(FXB1–FXB2+FXT1–FXT2)/1400.
$       +3.*L**2*(FYT1+FYT2)/(350.*HW)
$       +L**2*(FYB2+FYT2)/(60.*HW)
KG(5,29)=HW*L*(8.*FXB1–8.*FXB2+5.*(FXT1–FXT2))/12600.
$       +L**2*(FYT1+FYT2)/1400.
```

```
$       +L**2*(FYB2+FYT2)/300.
KG(6,6)=-L*(15.*FXB1-5.*FXB2-3.*FXT1+FXT2)/180.
$       +L**2*(FYT1+FYT2)/(105.*HW)
KG(6,9)=-(5.*FXB2-FXT2)/60.
$       +13.*L*(FYT1+FYT2)/(420.*HW)
KG(6,13)=L*(5.*FXB1-5.*FXB2-FXT1+FXT2)/360.
$       +L**2*(-FYT1-FYT2)/(140.*HW)
KG(6,17)=(FXB2+FXT2)/60.
$       +11.*L*(-FYT1-FYT2)/(210.*HW)
$       +2.*(L*FYB2+L*FYT2)/(20.*HW)
KG(6,21)=-L*(3.*FXB1-FXB2+3.*FXT1-FXT2)/180.
$       +L**2*(-FYT1-FYT2)/(105.*HW)
KG(6,24)=-(FXB2+FXT2)/60.
$       +13.*L*(-FYT1-FYT2)/(420.*HW)
$       +2.*(-L*FYB2-L*FYT2)/(20.*HW)
KG(6,28)=L*(FXB1-FXB2+FXT1-FXT2)/360.
$       +L**2*(FYT1+FYT2)/(140.*HW)
$       +2.*(L**2*FYB2+L**2*FYT2)/(120.*HW)
KG(7,7)=-HW**2*L*(21.*FXB1-7.*FXB2+3.*FXT1-FXT2)/12600.
$       +2.*HW*L**2*(FYT1+FYT2)/1575.
KG(7,10)=-HW*(23*FXB2-FXT2)/2100
$       +13.*L*(FYT1+FYT2)/4200.
$       +L*(FYB2+FYT2)/50.
KG(7,11)=-HW**2*(7.*FXB2+FXT2)/4200.
$       +13.*HW*L*(FYT1+FYT2)/3150.
KG(7,12)=-HW*L*(23.*FXB1-23.*FXB2-FXT1+FXT2)/12600.
$       +L**2*(FYT1+FYT2)/1400.
$       +L**2*(FYB2+FYT2)/300.
KG(7,14)=HW**2*L*(7.*FXB1-7.*FXB2+FXT1-FXT2)/25200.
$       +HW*L**2*(-FYT1-FYT2)/1050.
KG(7,18)=HW*(5.*FXB2+8.*FXT2)/2100.
$       +(-11.*L*FYT1-11.*L*FYT2)/2100.
$       +L*(FYB2+FYT2)/50.
KG(7,19)=-HW**2*(FXB2+FXT2)/1400.
$       +11.*HW*L*(-FYT1-FYT2)/6300.
$       +HW*L*(-FYB2-FYT2)/300.
KG(7,20)=HW*L*(15.*FXB1-5.*FXB2+8.*(3.*FXT1-FXT2))/6300.
$       +(L**2*FYT1+L**2*FYT2)/1050.
KG(7,22)=HW**2*L*(3.*FXB1-FXB2+3.*FXT1-FXT2)/4200.
$       +HW*L**2*(-FYT1-FYT2)/3150.
KG(7,25)=-HW*(5.*FXB2+8.*FXT2)/2100.
$       +(-13.*L*FYT1-13.*L*FYT2)/4200.
```

```
$       +L*(-FYB2-FYT2)/50.
KG(7,26)=HW**2*(FXB2+FXT2)/1400.
$       +13.*HW*L*(-FYT1-FYT2)/12600.
$       +HW*L*(FYB2+FYT2)/300.
KG(7,27)=-HW*L*(5.*FXB1-5.*FXB2+8.*(FXT1-FXT2))/12600.
$       +(-L**2*FYT1-L**2*FYT2)/1400.
$       +L**2*(-FYB2-FYT2)/300.
KG(7,29)=-HW**2*L*(FXB1-FXB2+FXT1-FXT2)/8400.
$       +HW*L**2*(FYT1+FYT2)/4200.
$       +HW*L**2*(-FYB2-FYT2)/1800.
KG(8,8)=-(15.*FXB1-55.*FXB2-3.*FXT1+11.*FXT2)/(36.*L)
$       +2.*(FYT1+FYT2)/(15.*HW)
$       +2.*(-FYB2-FYT2)/(4.*HW)
KG(8,15)=(5.*FXB1-15.*FXB2-FXT1+3.*FXT2)/(9.*L)
$       +(FYT1+FYT2)/(15.*HW)
KG(8,16)=-(FXB1-FXB2+FXT1-FXT2)/(36.*L)
$       +(FYT1+FYT2)/(30.*HW)
$       +2.*(-FYB2-FYT2)/(12.*HW)
KG(8,23)=-(3.*FXB1-11.*FXB2+3.*FXT1-11.*FXT2)/(36.*L)
$       +2.*(-FYT1-FYT2)/(15.*HW)
KG(8,30)=(FXB1-3.*FXB2+FXT1-3.*FXT2)/(9.*L)
$       +(-FYT1-FYT2)/(15.*HW)
$       +2.*(FYB2+FYT2)/(3.*HW)
KG(9,9)=-(5.*FXB1-5.*FXB2-FXT1+FXT2)/(10.*L)
$       +13.*(FYT1+FYT2)/(35.*HW)
$       +2.*(-FYB2-FYT2)/(4.*HW)
KG(9,13)=(5.*FXB1-FXT1)/60.
$       +11.*L*(-FYT1-FYT2)/(210.*HW)
KG(9,17)=(FXB1-FXB2+FXT1-FXT2)/(10.*L)
$       +9.*(-FYT1-FYT2)/(70.*HW)
$       +2.*(FYB2+FYT2)/(4.*HW)
KG(9,21)=-(FXB2+FXT2)/60.
$       +13.*L*(-FYT1-FYT2)/(420.*HW)
$       +2.*(L*FYB2+L*FYT2)/(20.*HW)
KG(9,24)=-(FXB1-FXB2+FXT1-FXT2)/(10.*L)
$       +13.*(-FYT1-FYT2)/(35.*HW)
KG(9,28)=(FXB1+FXT1)/60.
$       +11.*L*(FYT1+FYT2)/(210.*HW)
$       +2.*(-L*FYB2-L*FYT2)/(20.*HW)
KG(10,10)=-6.*(17.*FXB1-17.*FXB2-4.*(FXT1-FXT2))/(175.*L)
$       +78.*(FYT1+FYT2)/(175.*HW)
$       +(-FYB2-FYT2)/(2.*HW)
```

KG(10,11)=−HW*(23.*FXB1−23.*FXB2−FXT1+FXT2)/(350.*L)
$ +13.*(FYT1+FYT2)/350.
KG(10,12)=−(17.*FXB1−4.*FXT1)/175.
$ +11.*L*(FYT1+FYT2)/(175.*HW)
KG(10,14)=HW*(23.*FXB1−FXT1)/2100.
$ +11.*L*(−FYT1−FYT2)/2100.
$ +L*(−FYB2−FYT2)/50.
KG(10,18)=27.*(FXB1−FXB2+FXT1−FXT2)/(350.*L)
$ +27.*(−FYT1−FYT2)/(175.*HW)
$ +(FYB2+FYT2)/(2.*HW)
KG(10,19)=−HW*(8.*FXB1−8.*FXB2+5.*(FXT1−FXT2))/(350.*L)
$ +9.*(FYT1+FYT2)/700.
$ +(−FYB2−FYT2)/10.
KG(10,20)=9.*(FXB2+FXT2)/700.
$ +13.*L*(FYT1+FYT2)/(350.*HW)
$ +L*(−FYB2−FYT2)/(10.*HW)
KG(10,22)=HW*(8.*FXB2+5.*FXT2)/2100.
$ +13.*L*(FYT1+FYT2)/4200.
$ +L*(−FYB2−FYT2)/50.
KG(10,25)=−27.*(FXB1−FXB2+FXT1−FXT2)/(350.*L)
$ +78.*(−FYT1−FYT2)/(175.*HW)
KG(10,26)=HW*(8.*FXB1−8.*FXB2+5.*(FXT1−FXT2))/(350.*L)
$ +13.*(FYT1+FYT2)/350.
KG(10,27)=−9.*(FXB1+FXT1)/700.
$ +11.*L*(−FYT1−FYT2)/(175.*HW)
$ +L*(FYB2+FYT2)/(10.*HW)
KG(10,29)=−HW*(8.*FXB1+5.*FXT1)/2100.
$ +11.*L*(−FYT1−FYT2)/2100.
$ +L*(FYB2+FYT2)/50.
KG(11,11)=−HW**2*(7.*FXB1−7.*FXB2+FXT1−FXT2)/(700.*L)
$ +26.*HW*(FYT1+FYT2)/525.
KG(11,12)=−HW*(23.*FXB1−FXT1)/2100.
$ +11.*L*(FYT1+FYT2)/2100.
$ +L*(−FYB2−FYT2)/50.
KG(11,14)=HW**2*(7.*FXB1+FXT1)/4200.
$ +11.*HW*L*(−FYT1−FYT2)/1575.
KG(11,18)=HW*(5.*FXB1−5.*FXB2+8.*(FXT1−FXT2))/(350.*L)
$ +(−9.*FYT1−9.*FYT2)/700.
$ +(FYB2+FYT2)/10.
KG(11,19)=−3.*HW**2*(FXB1−FXB2+FXT1−FXT2)/(700.*L)
$ +3.*HW*(−FYT1−FYT2)/700.
$ +HW*(−FYB2−FYT2)/60.

KG(11,20)=HW*(5.*FXB2+8.*FXT2)/2100.
$ +13.*L*(FYT1+FYT2)/4200.
$ +L*(−FYB2−FYT2)/50.
KG(11,22)=HW**2*(FXB2+FXT2)/1400.
$ +13.*HW*L*(−FYT1−FYT2)/12600.
$ +HW*L*(−FYB2−FYT2)/300.
KG(11,25)=−HW*(5.*FXB1−5.*FXB2+8.*(FXT1−FXT2))/(350.*L)
$ +(−13.*FYT1−13.*FYT2)/350.
KG(11,26)=3.*HW**2*(FXB1−FXB2+FXT1−FXT2)/(700.*L)
$ +13.*HW*(−FYT1−FYT2)/1050.
KG(11,27)=−HW*(5.*FXB1+8.*FXT1)/2100.
$ +11.*L*(−FYT1−FYT2)/2100.
$ +L*(FYB2+FYT2)/50.
KG(11,29)=−HW**2*(FXB1+FXT1)/1400.
$ +11.*HW*L*(FYT1+FYT2)/6300.
$ +HW*L*(FYB2+FYT2)/300.
KG(12,12)=−L*(17.*FXB1−51.*FXB2−4.*(FXT1−3.*FXT2))/525.
$ +2.*L**2*(FYT1+FYT2)/(175.*HW)
KG(12,14)=HW*L*(23.*FXB1−69.*FXB2−FXT1+3.*FXT2)/6300.
$ +L**2*(−FYT1−FYT2)/1050.
KG(12,18)=9.*(FXB1+FXT1)/700.
$ +13.*L*(−FYT1−FYT2)/(350.*HW)
$ +L*(FYB2+FYT2)/(10.*HW)
KG(12,19)=−HW*(8.*FXB1+5.*FXT1)/2100.
$ +13.*L*(FYT1+FYT2)/4200.
$ +L*(−FYB2−FYT2)/50.
KG(12,20)=3.*L*(FXB1−FXB2+FXT1−FXT2)/1400.
$ +3.*L**2*(FYT1+FYT2)/(350.*HW)
$ +L**2*(−FYB2−FYT2)/(60.*HW)
KG(12,22)=HW*L*(8.*FXB1−8.*FXB2+5.*(FXT1−FXT2))/12600.
$ +L**2*(FYT1+FYT2)/1400.
$ +L**2*(−FYB2−FYT2)/300.
KG(12,25)=−9.*(FXB1+FXT1)/700.
$ +11.*L*(−FYT1−FYT2)/(175.*HW)
$ +L*(−FYB2−FYT2)/(10.*HW)
KG(12,26)=HW*(8.*FXB1+5.*FXT1)/2100.
$ +11.*L*(FYT1+FYT2)/2100.
$ +L*(FYB2+FYT2)/50.
KG(12,27)=−3.*L*(FXB1−3.*FXB2+FXT1−3.*FXT2)/700.
$ +2.*L**2*(−FYT1−FYT2)/(175.*HW)
KG(12,29)=−HW*L*(8.*FXB1−24.*FXB2+5.*(FXT1−3.*FXT2))/6300.
$ +L**2*(−FYT1−FYT2)/1050.

```
KG(13,13)=-L*(5.*FXB1-15.*FXB2-FXT1+3.*FXT2)/180.
$       +L**2*(FYT1+FYT2)/(105.*HW)
KG(13,17)=-(FXB1+FXT1)/60.
$       +13.*L*(FYT1+FYT2)/(420.*HW)
$       +2.*(-L*FYB2-L*FYT2)/(20.*HW)
KG(13,21)=L*(FXB1-FXB2+FXT1-FXT2)/360.
$       +L**2*(FYT1+FYT2)/(140.*HW)
$       +2.*(-L**2*FYB2-L**2*FYT2)/(120.*HW)
KG(13,24)=(FXB1+FXT1)/60.
$       +11.*L*(FYT1+FYT2)/(210.*HW)
$       +2.*(L*FYB2+L*FYT2)/(20.*HW)
KG(13,28)=-L*(FXB1-3.*FXB2+FXT1-3.*FXT2)/180.
$       +L**2*(-FYT1-FYT2)/(105.*HW)
KG(14,14)=-HW**2*L*(7.*FXB1-21.*FXB2+FXT1-3.*FXT2)/12600.
$       +2.*HW*L**2*(FYT1+FYT2)/1575.
KG(14,18)=-HW*(5.*FXB1+8.*FXT1)/2100.
$       +13.*L*(FYT1+FYT2)/4200.
$       +L*(-FYB2-FYT2)/50./
KG(14,19)=HW**2*(FXB1+FXT1)/1400.
$       +13.*HW*L*(FYT1+FYT2)/12600.
$       +HW*L*(FYB2+FYT2)/300.
KG(14,20)=-HW*L*(5.*FXB1-5.*FXB2+8.*(FXT1-FXT2))/12600.
$       +L**2*(-FYT1-FYT2)/1400.
$       +L**2*(FYB2+FYT2)/300.
KG(14,22)=-HW**2*L*(FXB1-FXB2+FXT1-FXT2)/8400.
$       +HW*L**2*(FYT1+FYT2)/4200.
$       +HW*L**2*(FYB2+FYT2)/1800.
KG(14,25)=HW*(5.*FXB1+8.*FXT1)/2100.
$       +11.*L*(FYT1+FYT2)/2100.
$       +L*(FYB2+FYT2)/50.
KG(14,26)=-HW**2*(FXB1+FXT1)/1400.
$       +11.*HW*L*(FYT1+FYT2)/6300.
$       +HW*L*(-FYB2-FYT2)/300.
KG(14,27)=HW*L*(5.*FXB1-15.*FXB2+8.*(FXT1-3.*FXT2))/6300.
$       +L**2*(FYT1+FYT2)/1050.
KG(14,29)=HW**2*L*(FXB1-3.*FXB2+FXT1-3.*FXT2)/4200.
$       +HW*L**2*(-FYT1-FYT2)/3150.
KG(15,15)=-4.*(5.*FXB1-5.*FXB2-FXT1+FXT2)/(9.*L)
$       +8.*(FYT1+FYT2)/(15.*HW)
KG(15,16)=(3.*FXB1-FXB2+3.*FXT1-FXT2)/(9.*L)
$       +(-FYT1-FYT2)/(15.*HW)
$       +2.*(FYB2+FYT2)/(3.*HW)
```

KG(15,23)=(FXB1–3.*FXB2+FXT1–3.*FXT2)/(9.*L)
$ +(–FYT1–FYT2)/(15.*HW)
$ +2.*(–FYB2–FYT2)/(3.*HW)
KG(15,30)=–4.*(FXB1–FXB2+FXT1–FXT2)/(9.*L)
$ +8.*(–FYT1–FYT2)/(15.*HW)
KG(16,16)=(11.*FXB1–3.*FXB2–5.*(11.*FXT1–3.*FXT2))/(36.*L)
$ +2.*(FYT1+FYT2)/(15.*HW)
$ +2.*(–FYB2–FYT2)/(4.*HW)
KG(16,23)=(FXB1–FXB2–5.*(FXT1–FXT2))/(36.*L)
$ +(–FYT1–FYT2)/(30.*HW)
KG(16,30)=–(3.*FXB1–FXB2–5.*(3.*FXT1–FXT2))/(9.*L)
$ +(FYT1+FYT2)/(15.*HW)
KG(17,17)=(FXB1–FXB2–5.*(FXT1–FXT2))/(10.*L)
$ +13.*(FYT1+FYT2)/(35.*HW)
$ +2.*(–FYB2–FYT2)/(4.*HW)
KG(17,21)=–(FXB2–5.*FXT2)/60.
$ +11.*L*(FYT1+FYT2)/(210.*HW)
KG(17,24)=–(FXB1–FXB2–5.*(FXT1–FXT2))/(10.*L)
$ +9.*(FYT1+FYT2)/(70.*HW)
KG(17,28)=(FXB1–5.*FXT1)/60.
$ +13.*L*(–FYT1–FYT2)/(420.*HW)
KG(18,18)=6.*(4.*FXB1–4.*FXB2–17.*(FXT1–FXT2))/(175.*L)
$ +78.*(FYT1+FYT2)/(175.*HW)
$ +(–FYB2–FYT2)/(2.*HW)
KG(18,19)=–HW*(FXB1–FXB2–23.*(FXT1–FXT2))/(350.*L)
$ +(–13.*FYT1–13.*FYT2)/350.
KG(18,20)=(4.*FXB2–17.*FXT2)/175.
$ +11.*L*(–FYT1–FYT2)/(175.*HW)
KG(18,22)=HW*(FXB2–23.*FXT2)/2100.
$ +11.*L*(–FYT1–FYT2)/2100.
$ +L*(–FYB2–FYT2)/50.
KG(18,25)=–6.*(4.*FXB1–4.*FXB2–17.*(FXT1–FXT2))/(175.*L)
$ +27.*(FYT1+FYT2)/(175.*HW)
KG(18,26)=HW*(FXB1–FXB2–23.*(FXT1–FXT2))/(350.*L)
$ +(–9.*FYT1–9.*FYT2)/700.
$ +(–FYB2–FYT2)/10.
KG(18,27)=–(4.*FXB1–17.*FXT1)/175.
$ +13.*L*(FYT1+FYT2)/(350.*HW)
KG(18,29)=–HW*(FXB1–23.*FXT1)/2100.
$ +13.*L*(FYT1+FYT2)/4200.
$ +L*(FYB2+FYT2)/50.
KG(19,19)=–HW**2*(FXB1–FXB2+7.*(FXT1–FXT2))/(700.*L)

```
$       +26.*HW*(FYT1+FYT2)/525.
KG(19,20)=-HW*(FXB2-23.*FXT2)/2100.
$       +11.*L*(FYT1+FYT2)/2100.
$       +L*(-FYB2-FYT2)/50.
KG(19,22)=HW**2*(FXB2+7.*FXT2)/4200.
$       +11.*HW*L*(FYT1+FYT2)/1575.
KG(19,25)=HW*(FXB1-FXB2-23.*(FXT1-FXT2))/(350.*L)
$       +9.*(-FYT1-FYT2)/700.
$       +(FYB2+FYT2)/10.
KG(19,26)=HW**2*(FXB1-FXB2+7.*(FXT1-FXT2))/(700.*L)
$       +3.*HW*(FYT1+FYT2)/175.
KG(19,27)=HW*(FXB1-23.*FXT1)/2100.
$       +13.*L*(-FYT1-FYT2)/4200.
$       +L*(FYB2+FYT2)/50.
KG(19,29)=-HW**2*(FXB1+7.*FXT1)/4200.
$       +13.*HW*L*(-FYT1-FYT2)/3150.
KG(20,20)=L*(12.*FXB1-4.*FXB2-17.*(3.*FXT1-FXT2))/525.
$       +2.*L**2*(FYT1+FYT2)/(175.*HW)
KG(20,22)=HW*L*(3.*FXB1-FXB2-23.*(3.*FXT1-FXT2))/6300.
$       +L**2*(FYT1+FYT2)/1050.
KG(20,25)=-(4.*FXB2-17.*FXT2)/175.
$       +13.*L*(-FYT1-FYT2)/(350.*HW)
KG(20,26)=HW*(FXB2-23.*FXT2)/2100.
$       +13.*L*(FYT1+FYT2)/4200.
$       +L*(FYB2+FYT2)/50.
KG(20,27)=-L*(4.*FXB1-4.*FXB2-17.*(FXT1-FXT2))/1050.
$       +3.*L**2*(-FYT1-FYT2)/(350.*HW)
KG(20,29)=-HW*L*(FXB1-FXB2-23.*(FXT1-FXT2))/12600.
$       +L**2*(-FYT1-FYT2)/1400.
$       +L**2*(-FYB2-FYT2)/300.
KG(21,21)=L*(3.*FXB1-FXB2-5.*(3.*FXT1-FXT2))/180.
$       +L**2*(FYT1+FYT2)/(105.*HW)
KG(21,24)=(FXB2-5.*FXT2)/60.
$       +13.*L*(FYT1+FYT2)/(420.*HW)
KG(21,28)=-L*(FXB1-FXB2-5.*(FXT1-FXT2))/360.
$       +L**2*(-FYT1-FYT2)/(140.*HW)
KG(22,22)=-HW**2*L*(3.*FXB1-FXB2+7.*(3.*FXT1-FXT2))/12600.
$       +2.*HW*L**2*(FYT1+FYT2)/1575.
KG(22,25)=-HW*(FXB2-23.*FXT2)/2100.
$       +13.*L*(-FYT1-FYT2)/4200.
$       +L*(FYB2+FYT2)/50.
KG(22,26)=-HW**2*(FXB2+7.*FXT2)/4200.
```

```
$       +13.*HW*L*(FYT1+FYT2)/3150.
KG(22,27)=-HW*L*(FXB1-FXB2-23.*(FXT1-FXT2))/12600.
$       +L**2*(-FYT1-FYT2)/1400.
$       +L**2*(FYB2+FYT2)/300.
KG(22,29)=HW**2*L*(FXB1-FXB2+7.*(FXT1-FXT2))/25200.
$       +HW*L**2*(-FYT1-FYT2)/1050.
KG(23,23)=(3.*FXB1-11.*FXB2-5.*(3.*FXT1-11.*FXT2))/(36.*L)
$       +2.*(FYT1+FYT2)/(15.*HW)
$       +2.*(FYB2+FYT2)/(4.*HW)
KG(23,30)=-(FXB1-3.*FXB2-5.*(FXT1-3.*FXT2))/(9.*L)
$       +(FYT1+FYT2)/(15.*HW)
KG(24,24)=(FXB1-FXB2-5.*(FXT1-FXT2))/(10.*L)
$       +13.*(FYT1+FYT2)/(35.*HW)
$       +2.*(FYB2+FYT2)/(4.*HW)
KG(24,28)=-(FXB1-5.*FXT1)/60.
$       +11.*L*(-FYT1-FYT2)/(210.*HW)
KG(25,25)=6.*(4.*FXB1-4.*FXB2-17.*(FXT1-FXT2))/(175.*L)
$       +78.*(FYT1+FYT2)/(175.*HW)
$       +(FYB2+FYT2)/(2.*HW)
KG(25,26)=-HW*(FXB1-FXB2-23.*(FXT1-FXT2))/(350.*L)
$       +13.*(-FYT1-FYT2)/350.
KG(25,27)=(4.*FXB1-17.*FXT1)/175.
$       +11.*L*(FYT1+FYT2)/(175.*HW)
KG(25,29)=HW*(FXB1-23.*FXT1)/2100.
$       +11.*L*(FYT1+FYT2)/2100.
$       +L*(-FYB2-FYT2)/50.
KG(26,26)=-HW**2*(FXB1-FXB2+7.*(FXT1-FXT2))/(700.*L)
$       +26.*HW*(FYT1+FYT2)/525.
KG(26,27)=-HW*(FXB1-23.*FXT1)/2100.
$       +11.*L*(-FYT1-FYT2)/2100.
$       +L*(-FYB2-FYT2)/50.
KG(26,29)=HW**2*(FXB1+7.*FXT1)/4200.
$       +11.*HW*L*(-FYT1-FYT2)/1575.
KG(27,27)=L*(4.*FXB1-12.*FXB2-17.*(FXT1-3.*FXT2))/525.
$       +2.*L**2*(FYT1+FYT2)/(175.*HW)
KG(27,29)=HW*L*(FXB1-3.*FXB2-23.*(FXT1-3.*FXT2))/6300.
$       +L**2*(FYT1+FYT2)/1050.
KG(28,28)=L*(FXB1-3.*FXB2-5.*(FXT1-3.*FXT2))/180.
$       +L**2*(FYT1+FYT2)/(105.*HW)
KG(29,29)=-HW**2*L*(FXB1-3.*FXB2+7.*(FXT1-3.*FXT2))/12600.
$       +2.*HW*L**2*(FYT1+FYT2)/1575.
KG(30,30)=4.*(FXB1-FXB2-5.*(FXT1-FXT2))/(9.*L)
```

$ +8.*(FYT1+FYT2)/(15.*HW)

where;

$$DW = \frac{Et^3}{12(1 - \nu^2)}, \qquad G = \frac{Et}{2(1 + \nu)}$$

$DD = Et, HW = h,$
$FXB1 = F_{x1}, FXB2 = F_{x2}, FXT1 = F_{x4}, FXT2 = F_{x5}$
$FYB2 = F_{y2}, FYT1 = F_{y4}$ and $FYT2 = F_{y5}.$

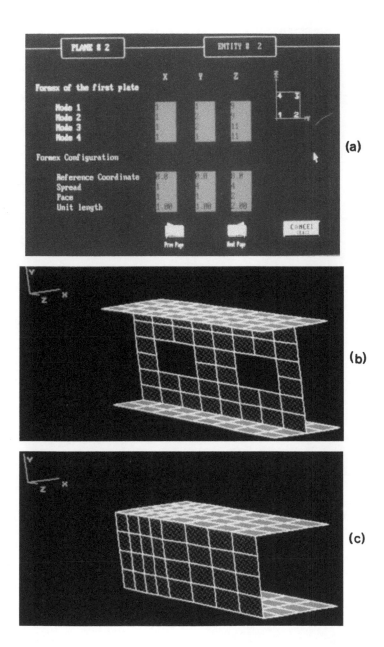

COLOR PLATE I. *See* F.G.A. Albermani and S. Kitipornchai, Figure 5, page 251.